Community Participatory Involvement

Community Participatory Involvement

A Sustainable Model for Global Public Health

Linda M. Whiteford
Cecilia Vindrola-Padros

Walnut Creek, California

LEFT COAST PRESS, INC.
1630 North Main Street, #400
Walnut Creek, CA 94596
www.LCoastPress.com

ISBN 978-1-62958-102-6 hardback
ISBN 978-1-62958-103-3 paperback
ISBN 978-1-62958-104-0 institutional eBook
ISBN 978-1-62958-105-7 consumer eBook

Library of Congress Cataloging-in-Publication Data on file

Printed in the United States of America

⊗™ The paper used in this publication meets the minimum requirements of American National Standard for Information Sciences—Permanence of Paper for Printed Library Materials, ANSI/NISO Z39.48–1992.

Contents

Illustrations

Figures

Tables

Preface

This book is written for people who want to make a difference and who want to help make the world a healthier place. Using the South American cholera epidemic as an example, the book demonstrates how a global community-based model known as Community Participation Involvement (CPI) successfully helped control, and finally stop, the spread of the disease. The model can easily be replicated for other diseases and applied in other cultural settings, and has a proven history of being sustainable. The book is designed to provide researchers, students, and practitioners with a brief introduction to the model and its tools, techniques, and methods.

In order to understand the conceptual basis of this model, we provide some history of its development, the theoretical and ethical issues involved in its application, and an example of the model in use. In-class experiential learning activities, a list of on-line resources and health related databases and a link to a ten-minute video from the research site (see Appendix D, tinyurl.com/WhitefordCPI) are included, and in addition, the book contains detailed information on the instruments and other tools developed for the cholera intervention and the project's associated workshops.

Readers will meet Ecuadorian people like Mariana and others whose lives were shaped by both the disease and the methods used to control it and learn how community members and public health practitioners like Isabel, Sofia, and Laura used the CPI model to shape their own futures. Our analysis draws on several streams of social science theory and public health models for interpreting the world in order to understand how the model fits within the current discourse around global health and infectious disease. Cultural practices, along with the biomedical disease paths, are considered together as necessary conceptual tools in the fight to control infectious disease. And they all come together through the active involvement of local communities whose members help to define parameters of research, the terms of engagement, the level of their participatory involvement, and the commitment to sustain the changes generated.

We want to acknowledge our appreciation for those visionaries in Ecuador and in the United States, like May Yacoob, whose commitment to health made this project possible, to the Ecuadorians – like Carmen Laspina, Mercedes Torres, Adela Vimos, and Nancy Benitiz – who passionately believed in this project and worked unfailingly to make it successful, and to the University of South Florida graduate students, Mika Kadono and Ryan Logan, and the external reviewers who worked with us to bring the story to publication.

Chapter One

Introduction

Mariana, a bright young woman in her late thirties, lived in the beautiful, isolated and rural highlands of Ecuador, surrounded by white-tipped volcanoes and mountain-fed streams. With its nestled communities and regional market centers, most indigenous people native to Ecuador live in these high mountains of the Andes – along with other native peoples living in equally beautiful and isolated areas in the lowlands of the Amazon. Some people living in the highland communities like Mariana had electricity, but most families did not have potable water or access to a sewage system. Most families had a pit latrine behind their house, while others practiced open defecation in the surrounding fields. The houses often had earthen floors and had neither glass nor screens on the windows. In the winter the houses were cold and damp, and in the summer the thin mountain light was clear outside. But inside the house, it was still dark.

When Mariana's father became ill, no one in the community knew what was making him sick. After he and several others from the community died, the doctor assigned to the regional health post came and talked

Community Participatory Involvement: A Sustainable Model for Global Public Health by Linda M. Whiteford and Cecilia Vindrola-Padros, 13–34.

to the members of the community. He said that what was killing them was called cholera and it was in their water.

> *"We didn't know what to do," Mariana said. "He said we should boil our water, but to do that we have to get extra wood or charcoal to burn, and that is expensive."*
>
> *"He told us to wash our hands, but to do that we need more water, and we don't have time to bring more water to the house."*
>
> *"He told us not to have our animals in the kitchen with us, but where are we going to put them at night to be safe if not with us?"*
>
> *"What could we do? We tried to ask him more questions, but he only spoke Spanish, not Quechua."*

Even with the best of intentions, the advice that Mariana and others in the community received was not helpful to them. Without intending to, they were immediately being 'non-compliant' with what the doctor told them to do. But according to Mariana, they had little choice. "If we had water in our houses, if we had extra wood or charcoal, or if we had another place to keep our animals, we would do what he said. But we live far away from the cities; we live with others like us who are not like the people in the cities. And we are poor."

Figure 1.1. Chimborazo Volcano

Mariana's story is not unique; people in many parts of the world live as Mariana and her family do. Living without access to a safe and reliable supply of water and without access to sewerage or other hygienic ways to dispose of daily human waste exposes people to many infectious and contagious diseases otherwise controlled. This book offers readers a chance to learn about a global health model that has been successfully applied on several continents and to a wide variety of health problems – for instance, water-borne diseases, environmental contamination, and rodent-borne diseases. This model helped stop the spread of cholera in Mariana's community and in other highland communities in the Andes. This book shows how a behavior change model can be successfully applied and the outcomes sustained under difficult and trying environmental conditions.

As we were writing this book, the world was shocked by the outbreak of another contagious disease. In this case it was the Ebola outbreak in West Africa. By December 2014, 5,987 people had died, 16,899 had been infected, and thousands of future cases were predicted in the coming months. Three West African countries (Guinea, Sierra Leone, and Liberia) were rapidly torn apart as they attempted to deal with the raging epidemic in the context of fragmented, understaffed, and resource-poor healthcare systems. The Ebola epidemic demonstrated to the world the consequences of failing to recognize the cultural and community-based components of infectious disease. The Ebola virus disease (EVD) has a high fatality rate and spreads through contact with the blood, secretions, organs or other bodily fluids such as in the vomit, mucus, breast milk, and semen of infected people. Infected people need to be identified rapidly and taken to facilities where they can be kept in isolation, and anyone whom they have been in contact with needs to be tracked and assessed for potential Ebola symptoms like fever, muscle pain, headache, sore throat, vomiting, diarrhea, rash, and bleeding. Anyone caring for a person infected with Ebola needs to wear protective gear to protect themself from infected bodily fluids. Clearly the role of the affected communities, cultural beliefs, and family practices needs to be understood.

The successful control of Ebola outbreaks requires a rapid response by identifying and isolating cases, providing adequate equipment for isolation and protective gear, and community engagement and education on how to care for the sick and bury those who have died. The death rate for women was significantly higher than that for men, suggesting the gender played an important role in patterns of exposure, and research demonstrated that the spread of the disease was linked to social and cultural factors that put women at a higher infection risk. In many parts of the world, women are the main caregivers and are thus in charge of caring for sick relatives, friends, and important people from their communities. Women are the ones in charge of washing and preparing bodies for burial. Women are also traditional healers and healthcare workers, roles and occupations that put women at a higher risk of infection because they are more likely to come into contact with the bodily fluids of infected people.

When we combine public health, epidemiology, and ethnographic research, we can see a nuanced and complex picture of the epidemic, one where the disease spreads in relation to traditional roles and cultural beliefs and practices. The recent Ebola epidemic, like the preceding cholera epidemic, demonstrates how critical knowledge of the local context is to understanding the distribution of cases and deaths and providing insight into how to control the epidemic. Twelve months after the Ebola epidemic was evident, the *Washington Post* (2014) published the following lessons: "rely on the local leadership, be sensitive to people's cultures, simple changes can yield significant results," each of which the global health model presented here does.

In this introductory chapter we identify some of the key theoretical and methodological components that underlie the Community Participatory Involvement (CPI) model. They will be discussed and presented in more detail in Chapter 3; here, we provide just a brief overview of key concepts. In this chapter we also include a discussion of leadership development as a component of the capacity building methodology at the heart of the CPI model. We also introduce readers to the disease of cholera and demonstrate how social analysis provides a critical lens for interpreting the course of the South American cholera epidemic. We conclude with a discussion of the ethical principles used in applied social science and public health research and, equally important, a discussion of the ethics of working in global, applied, and cross-cultural settings. While there are many academic programs that provide the pedagogical rationale for distinctive leadership techniques, here we offer an example from highland Ecuador, where the model was successfully applied during the cholera epidemic. Identifying potential leaders and providing space and the skills for them to develop those leadership capacities is a central concept in the CPI model, and the techniques are described here.

CPI Theory and Methods: Leadership Development and Capacity Building

The theory behind the CPI model is based on a medical ecology framework (Coreil 2010; McElroy and Townsend 2009) that considers medical vectors, geophysical surroundings, and cultural responses. While this theory is discussed in greater depth in Chapter 3, the use of the medical ecology framework helps explain the central role that leadership development and capacity building play in the CPI model.

Readers may remember the 2010 earthquake in Haiti, and the following cholera epidemic that killed more than 8,592 Haitians and sickened more than 706,089 (United Nations 2014). According to physician, anthropologist, and public health advocate, Paul Farmer, "It [was the] first big recrudescence of cholera in the Americas since the end of an epidemic that really swept through Peru and ended in 1993" (NPR 2011). Farmer continues, "If any country was a mine-shaft canary for the reintroduction of cholera, it was Haiti—and we knew it. And in retrospect, more should have been done to prepare for cholera ... which can spread like wildfire in Haiti. ... This was a big rebuke to all of us working in public health and health care in Haiti" (NPR 2011).

While there was some contention about the source of the introduction of cholera in Haiti following the earthquake, there is no question that leadership failed when it did not provide sufficient access to potable water and sewerage, and water-borne diseases spread like wildfire through the camps of the aid workers, peace keepers, and most of all, the homeless Haitians living in 'temporary' camps. The geography and ecology of Haiti made the spread of cholera quick and easy, and made the prevention of the disease slow and difficult. Haiti, like Ecuador, where our case study takes place, is a country interrupted by mountains, making transportation arduous and unreliable. The two countries share more than mountains, and where the cholera epidemic was most entrenched in the highlands of Ecuador, the two countries demonstrate a shared geopolitical reality of marginalized, rural, and dispersed populations living away from the urban centers. A shortage of medical personnel and health posts, absence of potable water and safe sewerage, lack of reliable transportation, and inadequate education created a similar pathway for the spread of cholera among the most impoverished rural families in both Haiti and Ecuador. The CPI medical ecology framework is well suited to understand and control cholera epidemics because of the powerful role that geography and ecology play in the distribution of this water-borne disease. This framework takes into account culture and history as prime drivers, and also considers the physical terrain and fauna.

Communities often are located along stream beds and in mountain valleys, thus making use of the abundance of water and the variable terrain of mountain sides for crops. This is certainly true in highland Ecuador and many other places. But the dispersal of villages and hamlets associated with such ecology often renders these same communities isolated from one another and in reduced contact with the administrative infrastructures of their governments. In addition to scarce contacts with governmental providers of health and education, remote highland communities often fail to maintain potable water and sewerage systems. As a result, they are exposed to water-borne and fecal-oral transmitted diseases.

Rural communities, whether in the United States or countries like Ecuador, often share characteristics related to their size, their degree of access to larger communities, presence of effective transportation routes for carrying goods and products to markets away from the communities, access to employment and education, and health care, for instance. Opportunities for leadership development, and institutional capacity building are likewise shaped by these characteristics. The CPI model is founded on the concepts of equity and equality, and those principles are translated into practice through leadership development. The process of community and participant selection (detailed in the case study in Chapter 5) is grounded in voluntary participation, autonomous decision-making, and equal opportunities to participate.

Taking into consideration local norms and customs, the categories of people traditionally under-represented in the area were given special consideration in order to equalize opportunities across the affected groups. In the highland Ecuador case, gender and age were two existing barriers to village leadership.

Even though the CPI model considers local norms, it also attempts to provide leadership opportunities to all members of the community who wish to participate. As a result, the CPI team decided to only work in communities that met several criteria: 1) the majority of the community leaders across a wide spectrum of formal and non-formal organizations voted to invite the project into their community, 2) leadership, as well as all other forms of participation, was not limited by categorical exclusions such as economics, gender, age, or nationality. As a result of these commitments, in several CPI communities women and young people were allowed to participate for the first time in village decision-making, with long-term positive results related to their sustained leadership. More details of the techniques and methods by which the leadership was engaged and trained can be found in Appendix A, "Brief Overview of Workshop Objectives, Contents, and Products," as well as described in Chapter 5.

Small villages or hamlets of several hundred families, often nestled into sheltered spaces or in open and relatively flat areas along the side of a mountain, face real and significant problems in trying to be connected to the world outside of their small area. In the United States, we can think of examples from Appalachia in which the hillside slopes are so intense that homes at the bottom are in shadows much of the day. Such steep slopes make transportation difficult, and even today with cell phones and computers, communication is difficult because of the service interruptions caused by the mountains. This kind of multi-faceted isolation restricts the number of options available to providers concerned with local capacity-building.

The CPI model overcomes some of these forms of isolation by developing institutional capacity both horizontally and vertically simultaneously. A core component of the model is to achieve agreement from governmental and non-governmental agencies, both in-country and outside, to support the project financially and with assigned personnel. This is a significant step because the in-country personnel provide the capacity to scale up and to create an institutional memory of the methodology and experience, while the foreign government financial support makes it possible to sustain the initial phases of the process.

In the Ecuadorian case, the Ministries of Social Well-Being and Health, Education, and Transportation all provided national and regional level representatives to commit work hours to the project for a period of twelve months. Representatives of several international non-governmental organizations (NGOs) also contributed personnel and financial support to the project. As a result, capacity building skills were transferred from the local to the regional, and then to the national level in these Ministries and NGOs as part of the CPI project.

Simultaneously, lateral contacts were made by members of various ministries with their counterparts in other ministries within the Ecuadorian government and with global NGOs. While the initial focus of the CPI project was to the stop the spread of cholera in identified communities of the high Andes, over a period of twelve months of working in the CPI project together, project members created new intra-ministry teams to work together on other problems. Regional representatives of education, for instance, found ways to assist their colleagues in health to augment health promotion campaigns through local schools and in teacher-education classes they were directing. Likewise, community members found themselves becoming friends with, and developing mutual interests and activities with, regional members of the various contributing agencies. In several cases, this resulted in villagers being helped to continue their education outside of the local village or to acquire job skills they could employ as representatives of the participating NGOs. In short, capacity-building became woven into the intra- and inter-team activities of the CPI project, extending skills beyond the scope of controlling the cholera epidemic.

Cultural Analysis of the South American Cholera Epidemic

Historically, cholera was referred to as the "Blue Death" because people turned blue as they died from the extreme loss of their bodily fluids. Today, the continued presence of the Blue Death is a conundrum: it still kills 100,000–130,000 people a year, yet now we know its cause, how to treat the disease, and that it can be easily prevented (Coupal 1995). Since John Snow isolated the mode of transmission through his experiment with the Broad Street pump in London in 1854, we've known how to interrupt the spread of cholera, and yet every day people die from the disease.

Cholera is an ancient disease caused by a bacteria, and its transmission is via a fecal-oral route usually associated with poor sanitation, such as people living without access to sanitary facilities or without a sufficient and reliable water supply. Public health researchers are interested in cholera not only because of its deadly properties, but also what its presence shows us about social conditions where it is found (Trostle 2005; Whiteford 2003). Cholera is found where the State is unable (or unwilling) to provide its citizens with a sanitary infrastructure of water and waste management, thus exposing them to preventable water-borne diseases (diseases produced by ingesting water that has been contaminated by human or animal excreta or urine containing pathogenic bacteria or viruses) (Montgomery and Eimelech 2007). Epidemiology – the study of the determinants and distribution of disease – allows us to track where outbreaks of the disease occur. By applying a medical ecology perspective, we learn why some populations are at greater risk for disease not due to biological causes, but due, instead, to patterns of social exclusion and disenfranchisement.

According to the U.S. Centers for Disease Control and Prevention (CDC), for the last 100 years, cholera has been almost non-existent in industrialized nations of North America and Western Europe. Yet the disease is still common today in other parts of the world, including the Indian subcontinent and sub-Saharan Africa. Latin America had not experienced an outbreak of cholera for close to a hundred years when in 1991 a cholera epidemic swept across Central and South America, killing thousands. Some suggest the re-emergence was due to changes in access/ingress of the cholera bacteria or vibrio (more ocean-going sea traffic for the waterborne bacterium to move in; more products being sold in various countries and transported long distances through air travel; more movement of people back and forth across borders). Others posit that it was a failure of national governments to provide a sanitary infrastructure to their growing urban populations (lack of water and sewerage treatment facilities) (Loyd 2009). Still other researchers suggested that once established, the distribution of the disease reflected the social structure of the populations where impoverished and stigmatized populations were most affected (Briggs 2001). There is, however, little doubt that the distribution of cholera in the South and Central American epidemic mirrored the social roots of poverty as it spread among vulnerable populations. Even more than twenty years after the South American cholera epidemic, cholera killed thousands in Haiti, once again following the social cleavages of poverty.

Social scientists, public health researchers, and practitioners concerned with social justice issues study the distribution of resources across society – access to housing, education, health care, employment opportunities – as a reflection of the social policies that protect or put at risk sectors of the population (Baer et al. 2003; Castro and Singer 2004; Farmer 1999, 2005; Schuller 2006; Singer et al. 1992). The presence of cholera signals vast social and economic disparities and, as a result, cholera is a fascinating disease to study, not only because of the social determinants of the disease, but also because of the behavioral interventions possible for interrupting its distribution.

The highlands of Ecuador and the lowlands of Venezuela, two regions in South America where the cholera epidemic spread, were ecologically distinct, but shared socio-cultural and economic similarities. The two affected regions shared conditions of poverty, marginalized indigenous populations, little access to electricity, few televisions sets or telephones for communication, few health posts or medical practitioners, and low levels of literacy (Briggs and Mantini-Briggs 2003; Whiteford et al. 1996). These characteristics made public health announcements delivered through mass media difficult, if not impossible. Rather, to combat the spread of the disease, public health messages had to be delivered person-to-person.

The public health campaigns to control the spread of the disease process were further complicated by a history of relations between the indigenous peoples and the national government weighted with a historical legacy of deceit and distrust. Anthropologist Charles Briggs and physician Clara Mantini-Briggs (2003) described some of the deadly consequences of that failure of trust during the cholera epidemic in Venezuela when the death rate increased because rural indigenous people delayed seeking medical help due to lack of trust in the authorities. Simultaneously, members of the Venezuelan media created negative and prejudicial representations of the indigenous people that exacerbated already existing tensions between the dominant (non-native) groups in power and the marginalized indigenous groups. What the Briggses termed the 'racialization of disease' epitomized the bias and prejudice that indigenous peoples experienced in Venezuela (and also in Ecuador), and resulted in their unwillingness to trust medical practitioners or others who appeared to be emissaries of those in power (such as most public health workers).

Argentina, like Venezuela, had a history of difficult relationships between those in power and indigenous peoples. Hugo Trinchero and Juan Martin Leguizamon poignantly identified this mistrust when they referred to "the established relationship between the treatment of the 'cholera question' and the 'indigenous question' in the mass media in Argentina" (1995:99). Their analysis of the local and national print media in Argentina demonstrated that the Argentine press was, similar to what the Briggses documented during the epidemic in Venezuela, reproducing discourses and ways of representing that stigmatized the indigenous groups, thus reproducing the symbolic violence in power relations. This mechanism of 'blaming the victim' for his or her own illness has the effect of distancing the national or regional government from the epidemic and re-directing the blame. Instead of indicting the government for the failure to provide a sanitary and safe infrastructure or to provide health care in a timely and appropriate fashion, the population itself is blamed, often for being indigenous, and their associated cultural practice indicted.

Trinchero and Leguizamon accused the media of re-contextualizing the cholera issue with specifically negative and prejudicial meanings, first blaming Bolivia (a country with a large indigenous population) for the entry of cholera into Argentina, then abandoning this "discussion of political frontiers" in favor of a focus on "cultural frontiers" (1995:103). The use of "easily decodificable stigmatized conceptions" like "'their culture', 'their hygiene habits', 'their migratory practices', etc." helped to make clear the line of blame the press was presenting, reflecting Bourdieu's concept of symbolic violence, as cited by Trinchero and Leguizamon (1995:104).

Indigenous groups were accused of being responsible for the distribution of cholera through their beliefs and behaviors, ignoring the practical realities of poverty and exclusion from resources that indigenous groups in Argentina and other parts of Latin America repeatedly experienced (Trinchero and Leguizamon 1995:104). According to Bourdieu (2001), when marginalized and dominated groups, like indigenous peoples in many countries, are subjected to the structural violence leveled through institutions of power, they sometimes internalize these discourses and blame themselves for their lack of access to resources and for the unequal distribution of disease. Thus, during both the Venezuela and Argentina cholera epidemics, rarely did the press identify the role that a history of prejudice and isolation played in the spread of the disease (Briggs 2008; Whiteford 2009a, b).

The reproduction of unconscious or cruel accusations targeting the poor or indigenous populations often occurs during times of crisis, particularly during the outbreak of new diseases. In their analysis of how poor, urban favela (slum) residents were depicted as causing the cholera epidemic in Brazil, Marilyn Nations and Cristina Monte (1996) documented how the victims were blamed. As with media accounts in both Venezuela and Argentina, the Brazilian governmental and municipal authorities were not initially indicted for their failure to provide for the safety of these citizens. Nations and Monte recorded what they referred to as the "metaphoric trappings" of racism in Brazilian public health messages that blamed the poor for the disease. However, the Brazilian favela residents were angered by the public health campaign that appeared to identify as them as the cause of the cholera epidemic. The public health message targeted residents and encouraged them to wash their hands after defecating, and identified the living conditions in the favelas as unsanitary. The campaign associated living "like a dog in dirt" with the outbreak of cholera (Nations and Monte 1996). In response, those living in the favelas pointed out that without water it is hard to wash, and without toilets, it is hard to manage waste. They argued that the presence of cholera in their communities was, instead, the fault of the government that failed to provide clean water and access to public sewerage in the favelas. Tragically, the social connotations of poverty, race, ethnicity, and the media-promoted image that they were 'dogs' kept people in the favelas from seeking immediate treatment for cholera because of their fear of further stigmatization, often dying as a result (Nations and Monte 1996).

These examples from Nations and Monte, Trinchero and Leguizamon, and Briggs and Mantini-Briggs show how vulnerable groups are blamed for the presence of disease in their communities, often without regard for the reasons that disease is there in the first place. These authors pull the viewer back from the daily lives of sufferers in order to bring the historical and political constraints into view, demonstrating the cruelty of blaming the victim for his or her poverty, disease, and even death. The gaze is shifted to the social history and institutional structures that repeatedly exclude, marginalize, and stigmatize people like the poor, in the Nation and Monte example, from the Brazilian slums.

As these researchers point out, one way to correct media misrepresentations of local people is through "community-based participatory research" so that community members have input into how public health messages are created (Briggs 2001). Briggs and Mantini-Briggs also suggest that community participation in the dissemination of public health information would facilitate the destruction of these pejorative images and would limit the spread of the dominant institutional models and practices, replacing them instead with models based on local realities (2003:313). As Chapter 6, "Outcomes and Discussion," and Chapter 7, "Reflections and Lessons Learned," show, the CPI is designed to engage community members in multiple transformations, including their ability to shape the messages made about them.

Ethics in Public Health Research and Practice

As part of the ethical considerations in writing this book, we kept all of the actual community names and geophysical locations; they can be found on maps. The names of the people in those communities and of those who participated in the CPI project are, however, pseudonyms. In addition to the role of theory and methods, ethics play a critical part in the design, delivery, and reception of any public health project. While the articulation of ethics may vary across time and place, and differ in various languages and cultures, ethical codes across time and space share many of the same underlying principles. Some of the principles reviewed here are those defined in the Belmont Report (Whiteford and Trotter 2008). The Belmont Report was a result of the National Commission for the Protection of Human Subjects involved in research in the United States (www.hhs.gov/ohrp/humansubjects/guidance/belmont.html). The Commission was initiated in response to the outrage over ethical failures such as the infamous Tuskegee experiment (Whiteford and Trotter 2008:33). We think these principles are powerful guides when working with vulnerable and/or culturally and linguistically distinct populations. These principles are:

1. Respect for persons;
2. Maximizing the good for people while minimizing any harm that may be done to them;
3. Insuring basic social justice for those people participating in research projects. (2008:46)

Concepts such as 'social justice,' 'respect,' 'good,' and 'harm' are complex and are embedded in the culture, history, and languages in which they are used. There is no single universal definition for these terms, but the underlying concepts can be found in cultures, religions, and language groups around the world.

Respect for Persons

Respect for Persons, the first principle, aims at protecting the rights of individuals over the rights of the State or government by assuring that individuals have the right to choose to participate, or not to participate, in a research project. Individual choice is the operant concept here, and it is the decision of the individual, not persons in authority over that individual, that is protected in this principle. For instance, individuals residing in State-run institutions may not be used as experimental research objects unless they knowingly choose to participate. Those who administer those institutions do not have the right to enroll residents in research projects without the consent of those individuals. This principle of respect for persons includes two separate ethical convictions. Individuals must be allowed to make decisions as autonomous actors for themselves, and those who are unable to make informed decisions for themselves, and who are thus considered a vulnerable population, must be protected. Institutionalized people, regardless of reason for their institutionalization (for instance, for being convicted of breaking the law, or being unable to protect themselves from harm) must be protected. Likewise, people who are incapacitated or immature (children) may require extensive protection. In the past, in the United States and in other countries, members of the armed forces were used for clinical trials for new and experimental vaccines. While this is no longer accepted practice, the Respect for Persons principle dictates that special considerations must be followed for such cases as a soldier's or prisoner's decision to participate in a clinical trial. This is to protect institutionalized people from coercion from those in authority. Whether the decision to 'volunteer' is the result of individual autonomy or the result of various forms of influence requires a consideration of the larger context in which it occurs, and it is often a matter of balancing competing claims stemming from the same principle. 'Respect for persons' could be interpreted to mean the need of protection, or on the other hand, to mean that the individual's decision (perhaps to volunteer) should be respected. The application of the ethical concept may stem from the identified principle, but may be used to support distinct, and sometimes conflicting, actions. Understanding the surrounding cultural norms and how a decision or application is contextualized is a critical element used to interpret these ethical principles in practice.

In addition to the issues of autonomy (the example of armed forces recruits 'volunteering' for experiments) and assessments of competence to participate, two further actions with which most public health researchers are familiar also derive from this first principle. They are the determination of 'Informed Consent' and the assertion of 'Confidentiality.' The questions raised are not only who might be allowed to participate in research, but how well the consequences of their participation are explained and understood, and who or what organizations/institutions might have access to the data generated by the research. Sometimes researchers, by the very nature of research, do not know what the results and repercussions of their research might be. Mixed methods research on the incidence of low-birth-weight babies in a particular hospital, for instance, might be designed to inform practice and policy aimed at reducing the numbers of underweight births. Postpartum women interviewed at their homes might provide their informed consents to engage in the project and be assured that the information they provided would be protected from individual identification. De-identified aggregate data are relatively safe from having individuals recognized. The narrative ethnographic component of the associated interviews, however, can expose information critical to understanding the cultural construction of pregnancy and pre-natal care and can contain stories that appear in research reports and presentations. Those narratives, even when the details and identities of the women are disguised, sometimes can be recognized by people who know them, and even by their providers.

Once published or presented publicly, the stories are out of the researcher's hands and, if picked up by other researchers or administrators, could cause unanticipated pain for the participants, even if none is intended. Thus, confidentiality can be assured, but not guaranteed.

Maximize Good and Minimize Harm

The second ethical principle is designed to maximize good for people and minimize harm. This same principle is sometimes referred to as the principle of beneficence (do no harm) and often identifies four stakeholder groups (subjects, communities, society, and science) that the researcher has an obligation to consider. The groups reflect the need to assess the risks and benefits in conducting any human research project and to be analytically able to move from a micro level of individual subjects and communities to the more macro level of society and science (Whiteford and Trotter 2008:73–74). Most social science research falls under the category of 'minimal risk,' as does most non-biological public health research. 'Minimal risk' means "the probability and magnitude of the harm or discomfort anticipated in the research are not greater in and of themselves than those ordinarily encountered in daily life or in the performance of routine physical or psychological examinations or tests" (Whiteford and Trotter 2008:74).

While the exact meaning of a concept such as 'harm' is difficult to assess, one understanding of the term is "actions that may result in physical, social, or psychological harm to the participants" (Whiteford and Trotter 2008:74). The U.S. National Institute of Health (NIH), a significant funder, and thereby assessor, of compliance with ethical guidelines, defines three levels of potential harm. In the category of 'physical harm,' the range of levels is from that of 'no harm,' to 'very slight harm' (such as experienced when getting a flu shot), to 'very severe harm,' and even death. 'Social harm' includes categories such as the loss of one's good reputation, disruption of family and kin relations, loss of job, or the creation of a stigmatized identity. In the category of 'psychological harm,' levels include transitory moments of, for instance, shame or embarrassment to traumatic and permanent psychological damage. Once again, the cultural context is critical in assessing the potential risk of different levels of harm and assessing the significance of that level of harm in its appropriate setting.

Ensure Social Justice

The third ethical principle to be considered is to ensure basic social justice for the people participating in human research. The principle of social justice presents the researcher with decisions about the selection of research subjects to make sure that each subject is considered for reasons directly related to the objectives of the research, and not for ease or availability or other secondary reasons. Minority status, age, gender, natal origins, and other variables can prejudice the selection of subjects and thereby both bias the outcomes of the research and also fail to consider the social justice of the selection. According to the Belmont Report: "Who ought to receive the benefits of research and bear its burdens? This is a question of justice, in the sense of 'fairness in distribution' or 'what is deserved.'" An injustice occurs when some benefit to which a person is entitled is denied without good reason or when some burden is unduly imposed (Whiteford and Trotter 2008:52, quoted from the Belmont Report 1979, part B: "Basic Ethical Principles").

Justice requires that the researcher attempt to protect all members of the project/intervention equally, regardless of individual attributes. A core aspect of the justice principle is that people have a right to expect fair and equitable treatment, and that the researcher is obligated to reduce/remove any untoward repercussion that may accrue to any one group or person. A particularly useful way to think about this principle is to consider the question of how to assure that no group of people will be exposed to the risks of research if their group will not also benefit from that exposure. That is, for instance, if the research protocol requires that all women in the sample be exposed to a toxic agent regardless of their age, but research on how to neutralize the toxin can be conducted only on pre-menopausal women, then the principle of justice is compromised because not all of those exposed can benefit from the potential results. They take the risks without any chance of the benefits; that is not considered just. As with all of these principles, there are cases in which the principles themselves come into conflict. In those cases, human decision-making must take into account the overarching questions of risks and benefits, and who and what benefits, as well as justifying the answer to why one group should benefit over another.

Global and Cross-Cultural Research

Ethical questions can be difficult to answer and are particularly complicated in cross-cultural situations. As all research becomes increasingly global and the changes in global systems such as transportation, communication, financial institutions, and human relations occur, the question of ethics becomes more complex. And also, oddly enough, it becomes easier. Cross-cultural research requires a sophisticated understanding of cultural institutions, historical and economic interdependency, and issues of scale, time and place. It also requires a respect and tolerance for differences and an ability to legitimate ideas contrary to one's own. Engaging in global research is, in and of itself, a complex and challenging endeavor that requires serious consideration of the primacy of distinct, and often competing, intellectual paradigms.

Review boards – for instance, the Institutional Review Boards (IRB) that are mandated in U.S. organizations, such as universities or other organizations that accept funding from the Federal government – do not exist in many other countries. Similarly, concepts central to our discussion here, such as 'respect,' 'confidentiality,' 'beneficence,' and 'justice,' may connote radically different meanings in distinct cultural settings. Indeed, the Belmont Report, from which we have drawn much of this presentation of ethics in research and practice, may not be known or valued.

For public health researchers one contrastive set with which we are most familiar is the artificial opposition of biomedical knowledge with traditional, non-biomedical, and perhaps indigenous conceptual frameworks. Unpacking this purported contrast allows us to see that these paradigms are different ways of interpreting the world; both are alternative bodies of knowledge that together constitute parts of our known worlds, neither complete by itself. That's the easy part. Once we recognize that our view of the world and, by definition, our ethical frameworks are situational and evolving, it provides greater justice to the engagement by leveling the playing field from one of absolute authority to one of collaboration and mutually constructed intelligence. The key terms from the Belmont Report presented in this chapter provide a set of guiding tools to be used with, and collaboratively redefined with, our global research partners.

U.S. researchers conducting projects outside of the United States should apply the same standards and protections as specified for research within the United States (www.hhs.gov/ohrp/international/). Increasingly, countries and organizations have their own review boards, and funders require that those review board sign off to approve research protocols. In the case described in this book, the project was not designed as research, but rather as a bi-lateral development and education project, focused on creating locally sustainable leadership and behavior changes. We did, however, seek and receive approval and buy-in at each stage of the project and with each level of stakeholder (local community people, politicians, NGOs, businesses, churches, regional, state, and finally national level representatives).

Mariana's story and the CPI model began this chapter. In the following chapters we review several important behavior change models and their successes and failures (Chapter 2); present the development of the CPI model, its theoretical framework, key concepts, and methodology in depth (Chapter 3). Chapter 4 provides the reader with the epidemiology of the South American cholera epidemic; Chapter 5 is the case study of the application of the CPI model to the epidemic. In Chapters 6 and 7, we present the quantitative outcome measures of the behavior changes that resulted from the CPI project, along with narratives from the individuals who participated, and finally, we conclude with a series of take-away messages and lessons learned from the application of the CPI model. Each chapter concludes with a brief summary of the chapter and several in-class exercises derived from the preceding materials.

Chapter Summary

- Global health, like cholera and Ebola, are world-wide.
- Global health requires an effective infrastructure and supplies in order to respond rapidly and effectively.
- Global health depends on effective leadership development and institutional capacity building programs.
- Global health requires the understanding of cultural beliefs and practices.
- Global health reflects the socio-cultural fault lines of poverty and prejudice.
- Global health depends on careful application of ethical principles.

In-class Exercise: Identifying the Cultural Context

Activity: Divide the group into teams of two or three people and give them 20 minutes to read the following essay about the cultural groups in the Ecuadorian highlands.

Goal: Participants learn to identify how cultural beliefs and practices, as well as history and economics, shape beliefs and practices related to health.

Method: Have each team identify and then present responses to the following exercises:

1. List at least three cultural attributes (beliefs and practices) identified in the essay that might shape the course of an epidemic.
2. Extract from the essay several cultural beliefs or practices that might shape the acceptance of public health/disease intervention practices.
3. Identify at least two ethical principles and show how their application might cause conflicts when applied cross-culturally in a behavior change intervention.

Essay: Cultural Context – Life in the Ecuadorian Highlands

> *"Life in the communities is very hard. The high altitude makes it cold and damp, with no heat or electricity. Even if they have electricity, the homes are dark and musty with animals and produce sharing the living spaces with the family."*

Isabel knew about rural life, but from afar. Isabel's own family came from the capital city, Quito, far away from the communities where cholera seemed unstoppable. Nevertheless, she understood how indigenous communities suffered exclusion from the mainstream of Ecuadorian life.

> *"They [indigenous communities] are proud of their language (Quechua), their traditions, and their traditional lifestyles, but they also suffer many losses that the rest of Ecuador doesn't – they have high infant mortality rates, few employment opportunities, and to be able to go on to school they have to move out of their communities."*

Ecuador is one of the smallest countries in South America, situated in the northwestern part of the continent between its larger and more powerful neighbors, Colombia and Peru. Ecuador crosses the equator (hence its name), lying in both the Northern and Southern Hemispheres. Geographically limited in space (260,000 sq km), it encompasses both extraordinary natural biodiversity of birds and plants and a rich cultural diversity. Indigenous cultures, such as the Shuar, Chachis, and Achuar, compose some of the more than fourteen identified and distinctive ethnic groups that live in Ecuador. Most travelers to Ecuador know the two primary cities: Quito, the Andean capital situated between the two cordilleras of volcanoes that create the "spine" running from north to south in Ecuador, and Guayaquil, the large coastal city on the Pacific side of the country. Other travelers know the Amazon in the south, or the famous weaving center, Otovalo, in the north of the country.

As a secondary center of the Spanish Empire, Ecuador never experienced the degree of glory or hardships of the Spanish colonization that Peru or Colombia did, but reflects many of the cultural and physical inheritances of the Spanish occupation in its people, customs and architecture. Along with the Spanish architecture of central plazas, overarching balconies, and internal patios, beliefs about European superiority were also transported. As a result, even in the middle of the twenty-first century, indigenous groups such as the Quechua are among the most economically deprived groups in the country. They live in remote areas with limited access to resources – including access to reliable and secure sources of water – as well as education and employment (Whiteford 2009b).

In urban Ecuador the cholera epidemic was relatively quickly controlled through a variety of traditional public health techniques, such as a massive health educational campaign to get citizens to wash their hands, to avoid eating food prepared and sold on the street, to avoid raw shellfish or other uncooked sea-food, and to use toilets, rather than empty lots and back alleys, for defecation. This campaign was designed to interrupt the major transmission pathways for cholera, which is spread through fecal–oral contact, or by not washing hands after coming into contact with contaminated feces. It is also spread when someone consumes uncooked seafood or other foods in which the vibrio is alive (from the water). The cultural practice of consuming traditional meals like ceviche (a meal of uncooked seafood like shrimp, mussels, and fish) increases the possibility of passing on the cholera vibrio. These two routes of cholera dissemination were identified early and targeted in urban centers, such as Quito and

Guayaquil, and although street venders and seafood sellers suffered from the results of the public health campaign, it did reduce the spread of cholera in the cities.

A second major thrust of the urban campaign to control cholera was to provide waste removal and public toilets. Previous to the cholera epidemic, in urban centers such as Quito it was close to impossible to find a public toilet. As a result, people had little recourse other than to use unoccupied spaces (back alleys, empty lots, and so on). The interventions were classic public health interventions, such as disease education, provision of safe or uncontaminated water, and proper sanitation, and they were successful in interrupting the transmission of cholera in the urban areas.

Within thirteen to eighteen months from the inception of the cholera epidemic in Ecuador, the numbers of confirmed cases in Quito began to decline. But in the rural highlands of the Andes, far away from the modern public infrastructure of the urban centers, cholera continued to spread with deadly consequences where Mariana and her family lived.

Chapter Two

Global Health and Behavior Change Interventions

The design and implementation of health interventions is a long and complex process involving multiple factors and stakeholders. As Panter-Brick and colleagues have argued, there are only a few examples of "truly successful" health interventions (2006:2811). The success of health interventions depends on the underlying ideas guiding the intervention (often called theories of change), financial and political support, the ways in which the intervention is delivered (often called implementation), the engagement of relevant stakeholders along all stages of design and implementation, the resources available to guarantee the maintenance of the intervention over time (sustainability), and the relevance and appropriateness of the intervention (considering the cultural, social, and economic context where it is applied), among many others.

An intervention that works well in one place might not work at all in a different context. An intervention that looks good on paper might not function in practice. An intervention some people think is excellent might not be viewed positively by others. In this chapter we will provide

Community Participatory Involvement: A Sustainable Model for Global Public Health
by Linda M. Whiteford and Cecilia Vindrola-Padros, 35–49.

an overview of health interventions, the main models used to guide them, proposals for planning and implementation, and designs for evaluations. We present general guidelines but recognize that each intervention will be unique in the sense that it will need to adapt to specific contexts, target populations, resources, and barriers.

Behavior Change Models

When designing any health intervention, designers are forced to grapple with theories on behavior change. Health interventions are based implicitly or explicitly on one or multiple behavior change models. Even those interventions that do not acknowledge using a particular model for understanding how changes in behavior will be made have underlying ideas about the factors that influence individual decisions and actions and how these affect their health. Therefore, in this section we provide a brief overview of the main behavior change models used in health interventions.

Different types of behavior change might be required to deliver an intervention. People delivering the intervention might need to undergo some degree of behavior change, such as healthcare professionals who need to incorporate new guidelines into their practice to provide better care to patients. The intervention might also be aimed at changing the behavior of those receiving it, such as a people who need to stop smoking, incorporate new hygiene habits, or increase their exercise routines to improve their health. This is the main reason why a significant amount of work around the design of interventions has focused on understanding the factors that determine behavior change. Several theoretical models have been proposed, and we have summarized them in Table 2.1.

Table 2.1. Behavior Change Model

Model name	Type of model
Health Belief Model (HBM)	Motivational model
Protective Motivation Theory (PMT)	Motivational model
Theory of Planned Behavior (TPB)	Motivational model
Information-motivation-behavioral skills model (IMB)	Behavioral enaction model
Transtheoretical model	Multi-stage model
Social ecological models	Social/environmental model

Health Belief Model

Motivational models are based on the idea that they can predict behavior. They focus on identifying the variables that influence the decisions people make about their health (Armitage and Conner 2000:174). A type of motivational model is the Health Belief Model (HBM), which states that health behavior is determined by the ways in which the individual perceives susceptivity (if they are susceptible to a negative health outcome), severity or threat, benefits, barriers; if they are motivated to engage in the behavior; and if there are any cues or hints that they need to take action (for instance, if they feel the symptoms of a disease or a convincing message is communicated through mass media campaigns) (Rosenstock 1966). Although widely used, the HBM has received significant criticism due to its inconsistency, inability to actually predict behavior, and the limitations produced by only focusing on four main variables (susceptivity, severity, benefits, and barriers) (Carpenter 2010).

A recent application of the HBM can be found in the work carried out by Walker and Jackson (2015) on the oral hygiene beliefs and practices of preteen children. Their study sought to understand why children practice good oral hygiene. They carried out a series of focus groups with children aimed at documenting their perception of good and bad oral hygiene, susceptibility for poor oral health, and barriers for practicing oral hygiene practices. They were able to identify two main factors that could act as potential motivators to engage in good oral hygiene practices: the esthetic appearance of teeth and the desire to please others by brushing (Walker and Jackson 2015).

Protection Motivation Theory

The Protection Motivation Theory (PMT) is another example of a motivational model. According to this theory, health behavior can be explained in terms of responses to a health threat that can be adaptive (beneficial to health) or maladaptive (harmful to health). The individuals make an appraisal of both the threat and the coping mechanisms at their disposal and perform an either adaptive or maladaptive response. An adaptive response leads to protection motivation, or "the intention to perform the health-protective behavior or avoid the health-compromising behavior" (Conner and Norman 2005:9).

Williams and colleagues (2015) have recently used this model to understand behavior in infectious disease epidemics. Their study focused on using PMT to determine if individuals would use social distancing behavior (the reduction of social contacts) in a simulated infectious disease epidemic. Research participants were asked to fill out a questionnaire and participate in a simulated epidemic in a computer game. The authors found that PMT could be used to explain participants' intention to engage in social distancing, but not their actual distancing behavior during the simulated epidemic (Williams et al. 2015).

Theory of Planned Behavior

The theory of planned behavior (TPB) stems from the theory of reasoned action (TRA), which maintains that intention is a "proximal determinant of behavior" (Armitage and Conner 2000:177). In TRA, intention is understood as the motivation an individual requires to carry out a particular behavior (Armitage and Conner 2000:177). TPB represents an extension of TRA because it goes beyond a vision of behavior as just being dependent on intention. It includes an analysis of the perceived behavioral control (PBC), where if the individual perceives he or she has greater control over the situation, the person is more likely to perform the behavior (Hardeman et al. 2002). In doing so, TPB considers the situations where individual intention to change the behavior is present, but the individual does not have control to change it (Armitage and Conner 2001:472).

An example of the use of this model is the study on healthy eating carried out by Conner and colleagues (2002). In this study, healthy eating behavior was predicted from intentions collected from research participants in different time intervals over six years. TPB explained 43 percent of the variance in behavioral intentions in specific time points and 20 percent over time (Conner et al. 2002). The findings led the authors to propose that healthy eating interventions need to focus on promoting intention or motivations to engage with healthier eating habits and to provide more information on how to translate these motivations into actual dietary changes (Conner et al. 2002:199). TPB has been recognized as effective in predicting behavior (Armitage and Conner 2001), but it has been critiqued due to the failure to translate it into the actual development and evaluation of interventions (Hardeman et al. 2002).

Information-Motivation-Behavioral Skills Model

The information-motivation-behavioral skills (IMB) model is based on the idea that relevant information, personal (individual attitudes) and social (perceived social support) motivation, and the skills required for performing a behavior are all important variables in determining if a behavior will be enacted. This model has been used to understand adherence to antiretroviral therapies by HIV positive patients. According to Fisher et al., "to the extent that HIV-positive individuals are well-informed, motivated to act, and possess the behavioral skills required to act effectively, they will be more likely to adhere to HAART [highly active antiretroviral therapy] regimens over time and to reap the substantial health benefits of this treatment" (2006:463). This model has also been used to predict breast self-examination in women, and study findings showed that lack of information, motivation, and behavioral skills led to low rates of breast self-examination (Misovich et al. 2003).

Transtheoretical Model

Multi-stage behavior change models are based on the idea that individuals go through a series of stages to change their behavior. The most commonly used multi-stage model is the transtheoretical model of change proposed by Prochaska and DiClemente (1994) (other examples of multi-stage models include the health action process approach, the Rubicon model, and Kuhl's action control theory). In the transtheoretical model each stage represents a different phase of motivational readiness for change. The model begins with a precontemplation stage where the person does not intend to make a change. The person then moves to the contemplation stage where change is usually intended within a six-month window. The preparation stage follows where the person actively plans and makes arrangements for change. In the action stage, changes are overtly made, and these are sustained during the maintenance stage (Prochaska et al. 1994).

An example of the application of some of the concepts of the transtheoretical model is Lippke and Ziegelmann's (2006) use of the multi-stage

Figure 2.1. Stages in the Transtheoretical Model of Behavior Change

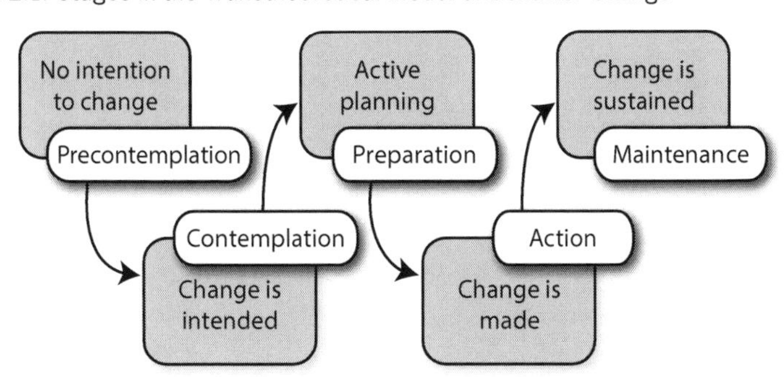

model (MSM) of changed behavior to understand the rehabilitation process of patients with orthopedic health limitations. Patients were asked about their engagement in physical exercise, and their responses were organized according to the different stages of the model (if they said no: precontemplation, contemplation, disposition, preaction; if they said yes: implementation, habituation, fluctuation). The authors then measured the model's capacity for predicting behavior changes in physical exercise (Lippke and Ziegelmann 2006).

Social Ecological Model

Most of the behavioral models discussed so far have emphasized the role of individuals rather than their environments in the process of behavior change. The Socio-ecological model seeks to expand our understanding of behavior change by looking at the dynamic interaction of individuals with their surroundings (Stokols 1992). Under this model, behavior is seen as determined by:

1. Intrapersonal factors (attitudes, knowledge, skills);
2. Interpersonal processes (formal and informal social networks, family, community groups);
3. Institutional factors (social institutions);
4. Community factors (relationships between organizations and institutions);
5. Public policy (local, state, and national laws and regulations).

(McLeroy et al. 1988:355)

When this model is used to understand behaviors related to health, it privileges the study of the multidimensional and complex qualities of the social and ecological environments where individuals interact on a daily basis (Stokols 1992). The model looks at factors affecting health at multiple levels (individual, groups, communities, regions, countries, etc.) and seeks to understand the interaction of different sectors (economic, political, cultural, etc.) (Glanz and Bishop 2010:403).

The social ecology model of health has been adapted in various ways. For instance, Coreil et al. (2000) have proposed the study of "the household ecology of disease transmission" to understand the role played by household variables in the transmission of dengue fever. According to the authors, by focusing on the physical setting where individuals carry out daily activities, such as eating, sleeping, and washing, it is possible to identify the risk and behaviors that determine disease transmission as well as those that might protect individuals from the disease (Coreil et al. 2000). By shifting the focus from the individual to the entire household, this model is able to capture "the interstices among the biophysical, social and culturally constructed environments as they shape health practices" (Coreil et al. 2000:168).

Panter-Brick and colleagues (2006) have also proposed using a social ecology model of behavior change to understand behavior in particular physical and social settings. According to the authors, this process of contextualization leads to the design of culturally acceptable interventions, where both intention and ability to change are taken into consideration (Panter-Brick et al. 2006). This is one of the main contributions of the social ecology model, as behavior change is not just seen as resulting from individual decisions to make the change (intention), but is understood as a complex process heavily influenced by the barriers they might encounter while attempting to make the change (ability). In order to combine these two dimensions of behavior change, the authors propose developing culturally compelling interventions – that is, interventions that "must be embedded in social and ecological settings, triggered by compelling messages for change, and explicitly evaluated in terms of perceived, objective and sustainable impact" (Panter-Brick et al. 2006:2813). The focus on culturally compelling interventions brings to light the need to design interventions in relation to local needs and priorities and the creation of a sense of community ownership of the intervention from initial stages of design (Panter-Brick et al. 2006:2812).

Community-based Participatory Models

The development of a wide range of collaborative approaches in public health has been linked to the emergence of social and ecological models and a critical reflection of the power differentials involved in the research process (Israel et al. 1998:175). Different labels, such as "participatory research," "action research," "cooperative inquiry," "empowerment evaluation," and "collaborative research" have been used to describe research centered on the development of partnerships between multiple groups of stakeholders and the involvement of participants in various aspects of research (Israel et al. 1998). Community-based participatory research (CBPR) has been proposed as a term that could potentially envelop several of these approaches (Minkler 2004). CBPR has been defined in different ways, with the variation in definitions often linked to different

Table 2.2. Some Definitions of CBPR

Author	Definition
Green et al. (1995:12)	"A systematic investigation with the participation of those affected by an issue for purposes of education, an action, or affecting social change"
Rhodes et al. (2010:174)	"Establishes structures for full and equal participation in research by community members (including those affected by the issue being studied), organizational representatives, and academic researchers to improve community health and wellbeing through multilevel action, including individual, group, community, policy, and social change"
Wallerstein and Duran (2006:372)	"An orientation to research that focuses on relationships between academic and community partners, with principles of colearning, mutual benefit, and long-term commitment, and incorporates community theories, participation, and practices into the research efforts"

interpretations of its main components: participation, research, and action (Minkler 2004:ii3). We have outlined some of these definitions in Table 2.2.

Even though the application of CBPR in practice might vary depending on the purpose of the research and the context, a series of key principles of this approach have been identified. CBPR recognizes the community as a unit of identity; it builds on the community's strengths and resources; facilitates collaborative partnerships, co-learning, and the co-production of knowledge; entails capacity building of community members; integrates knowledge and action for the benefit of all partners; involves iterative processes; and implements open and inclusive dissemination strategies (Israel et al. 1998; Minkler 2004; Wallerstein and Duran 2010).

These key principles help explain the main benefits of using CBPR approaches. By ensuring this level of involvement of community groups, CBPR often leads to the design of research that reflects the real concerns and needs of community members. Engagement of community members in the research process might be achieved faster and easier because the research is in their best interest and they have experienced a sense of ownership of the project (due to their early participation in its development) (Lee et al. 2003). The active inclusion of community members in research can also lead to the uncovering of important lay knowledge on the topic that might otherwise not have been available to the researchers (Hall 1992). In cases where CBPR is used to implement interventions, active community involvement can increase the relevance of the intervention, its success, and long-term sustainability (Schulz et al. 1998; Wallerstein and Duran 2010).

Real and sustained community engagement and participation are not easy to achieve, and researchers often face barriers and challenges unique to this approach. Lack of trust is encountered frequently and can be produced by the inability of researchers to gain trustworthiness as well as previous negative experiences or fears that might be deterring community members from getting involved (DeKoning and Martin 1996). Another challenge is related to the representativeness of community-based approaches and the fact that communities are not homogeneous (Israel et al. 1998). Power differentials within communities might lead to the exclusion and, hence, lack of participation of marginalized members or groups (Altman 1995). Once community members are brought together, it might be difficult for them to reach common goals, and conflicts between different priorities and interests can limit their degree of participation. Furthermore, community-based approaches usually entail a significant investment in time and money (Israel et al. 1998).

The Community Participatory Involvement (CPI) model, which we will describe in greater detail in Chapter 3, is a variation of community-based models. The CPI model differs from other community-based participation models in that its focus lies in the relationships community members are able to establish with different sectors of the state and civil society. In this model, even though the engagement of community members is seen as intrinsic in the generation of changes, these changes are seen as requiring input and support from a wide range of stakeholders. Local community needs are seen as the responsibility of community members, different levels of political and civil authority, and non-governmental organizations.

Implementing Health Interventions

The implementation of health interventions is complex and highly dependent on the context where the intervention will take place, the target population, the characteristics of the implementers, the purpose of the intervention, and the processes used to design the intervention (Guttmacher et al. 2010). Different models have been proposed to implement health interventions, but these should be taken as general guidance that will probably need to be adapted to suit local realities, interests, and demands.

Bartholomew et al. (2011) have proposed the use of intervention mapping techniques to guide intervention design and implementation. Intervention mapping is based on the following steps:

1. Needs assessment: engage with different stakeholders to determine the needs of the target population, assess capacity, and specify intervention goals
2. Matrices: state desired outcomes and performance objectives (how the outcomes will be measured)
3. Theory-based intervention methods and practical applications: identify suitable behavior change models, develop program methods
4. The intervention program: draft and pretest or pilot program materials and methods
5. Adoption and implementation: set up and implement the intervention
6. Evaluation plan: design, pilot, and implement the evaluation

(Bartholomew et al. 2011)

Ideally, the intervention implementers would engage with relevant stakeholders throughout all stages of the intervention (from the initial needs assessment to the final evaluation) to find out their views and incorporate their knowledge. Local views on health and illness are crucial for any intervention as they "flag local health concerns" (Nichter 2008:69). As Nichter (2008) has argued, local views on illness provide insight into: the symptoms that are given importance or cultural significance, illnesses considered serious vs. those seen as mundane, how views on illness and their classification influence treatment seeking behaviors.

Evaluating Health Interventions

Evaluations of interventions are usually performed to determine the effectiveness, value, and sustainability of the changes (in health, service delivery, process, etc.) produced by the intervention (Sidani and Braden 2011). Evaluations can be carried out simultaneously with the intervention to help shape it along the way (known as formative evaluations) and ensure it is better suited to address the needs of the target population, or they can be done once the intervention is completed to assess its performance and effectiveness and, potentially, inform future work (known as summative evaluations) (Stetler et al. 2006). According to Michie and Abraham (2004), evaluations of behavior change interventions should be able to answer the following questions:

1. Does it work?
2. How well does it work?
3. How does it work?

When we ask if an intervention works, what we are mainly interested in finding out is if the intervention has produced an improvement. We basically want to know if what was deemed a potential improvement during the design phase has become an actual improvement in practice. Measuring improvement is challenging and seldom straightforward. What might represent an improvement for some groups might not be beneficial for others. The intervention might produce improvements in health outcomes, but it might elevate healthcare costs, thus pointing to potential issues in sustainability (Ovretveit 2014:5).

When we look at how well an intervention works, we are judging its value (Michie and Abraham 2004). This judgment can be made solely by the evaluator, or, in what is now a more common strategy, it can be reached through the collaborative work of the evaluator with different groups of stakeholders (implementers of the intervention, people receiving the intervention, the community, patient and public involvement groups, etc.).

This value judgment can be informed by an evaluation of how the intervention works. This is normally called a process evaluation, where the evaluator analyzes the mechanisms, procedures, and strategies used to design and implement interventions (Hulscher et al. 2003). Process evaluations facilitate the replicability of the intervention, but they also help explain why an evaluation works or does not work and how well it works. Process evaluations are based on the idea that the intervention design or blueprint might not be implemented in the same way across all contexts or with all groups (Ovretveit 2014; Saksvik et al. 2002). By looking at how the intervention is "performed," process evaluations point to the variability in implementation and help identify factors that might have hindered the true potential of the intervention for making health improvements (Hulscher et al. 2003).

Evaluations of health interventions can be designed in different ways depending on the purpose of the evaluation, research questions, and the context where the intervention was implemented (Ovretveit 2014; Wynn et al. 2006). We have included a summary of designs in Table 2.3. Observational designs focus on understanding the changes produced by interventions in their actual context. Experimental and quasi-experimental designs view health interventions as experiments with measurable outcomes that can be predicted before the intervention is implemented. Action evaluation designs are characterized by the active involvement of the evaluator in the shaping and improvement of the intervention (Ovretveit 2014; Wynn et al. 2006).

Once the evaluation has been designed, then the practical issues, such as data collection, analysis, and dissemination, need to be resolved.

Table 2.3. Types of Intervention Evaluation Designs

Observational designs	Audit evaluations
	Cohort evaluations
	Cross-sectional studies and case-control studies
	Qualitative or mixed-methods observational evaluation designs
	Single case evaluation
Experimental and quasi-experimental designs	Before/after design
	Time series design
	Quality improvement testing (PDSA)
	Comparative experimental designs
	Trials: randomized controlled trials (RCTs), non-randomized controlled trials, cross-over comparative trial, and stepped-wedge trial
Action evaluation designs	Formative evaluations
	Evidence-based quality improvement (EBQI)

(Sources: Ovretveit 2014; Wynn et al. 2006)

The type and quantity of data will depend on the purpose of the evaluation and its design, but experienced authors in evaluations, such as Ovretveit, have put together Ten Golden Rules for collecting and evaluating data for evaluations – rules which can apply to all of the designs we discussed above:

1. Don't collect data unless you are sure no one else has done so already
2. Don't invent a new measure when a proven one will do
3. When the person or documents you need are not there, don't be blind to what is there which could help – be opportunistic
4. Measure what is important, not what is easy to measure
5. Don't collect data where confounders will make interpretation impossible
6. Spend twice as much time on planning and designing the evaluation than you spend on data collection
7. Always do a small pilot to test the method on a small sample
8. As you collect the data, save them to a database which is designed with thought to how to carry out the analysis, and back it up every day!
9. Analyzing the data takes twice as long as collecting it, if you have not defined clearly which data you need and why
10. Remember, data collection will always take twice as long as you expect.

(Ovretveit 2014:37)

The strategies used for dissemination will also depend on the purpose of the evaluation, but results are usually shared with all of the relevant stakeholders. The type and content of the reporting will vary in relation to requirements from the funders, the interests and needs of the implementers and users of the intervention, and the context where both the intervention and evaluation took place. There are general guidelines for reporting findings that can be found in Boutron et al. (2008) (for randomized controlled trials), Ogrinc et al. (2008) (quality improvement), Thomas (2006) (evaluations using qualitative data), and Husereau et al. (2013) (for economic evaluations).

Chapter Summary

- The design and implementation of health interventions is a long and complex process involving multiple factors and stakeholders.
- Most interventions depend on some degree of behavior change and often use one of the following behavior change models: Health Belief Model, Protective Motivation Theory, Theory of Planned Behavior, Information-motivation-behavioral Skills Model, Transtheoretical Model, or Social Ecological Model.
- The implementation of health interventions is complex and highly dependent on the context where the intervention will take place, the target population, the characteristics of the implementers, the purpose of the intervention, and the processes used to design the intervention.
- Ideally, the intervention implementers would engage with relevant stakeholders throughout all stages of the intervention (from the initial needs assessment to the final evaluation) to find out their views and incorporate their knowledge.
- Evaluations of interventions are usually performed to determine the effectiveness, value, and sustainability of the changes (in health, service delivery, process, etc.) produced by the intervention.
- Three main approaches are used to design the evaluation of health interventions: experimental and quasi-experimental designs, observational designs, and action evaluation designs.

In-class Exercises: Implementing Behavior Change Models

Activity: Divide the class into small groups and have them review and discuss the various behavior change models presented in the chapter.

Goal: Class participants should be able to identify how various models differ from one another, and how to select the most useful model for their project.

Method: Open the discussion of the preferred models the groups in the class have identified and the justification for their selection by responding to the following guides:

1. Select one of the behavior change models presented in this chapter and describe how you would use it to design and implement an intervention.
2. Explain how you could involve community members throughout different stages of planning and implementation of the intervention.

Identify some of the obstacles you might anticipate in implementing a behavior change intervention.

Global Health Model

Community Participatory Involvement (CPI) Approach to Social Change

History and Applications of the CPI Model

Global health reflects the complex interplay of history, climate, politics, geography, and culture. And into this already complicated scenario, we add new and evolving diseases. The question we attempt to answer here is: How can one global health model help change behaviors with such a varied set of parameters? As the previous chapter discussed, there are many examples of behavior change projects that have been tried, with different levels of success and uneven applications. The Community Participatory Involvement (CPI) model we describe in this and subsequent chapters has had success in tackling global health problems ranging from environmental pollution, lack of community involvement, absence of appropriate institutional infrastructure, and housing and sanitation, to water-borne disease outbreaks in both geographic and cultural areas as distinct as the Middle East, Africa, and South America.

The model was developed and funded by the United States Agency for International Development (USAID) and was implemented by the

Community Participatory Involvement: A Sustainable Model for Global Public Health
by Linda M. Whiteford and Cecilia Vindrola-Padros, 51–70.

Environmental Health Project (EHP). Originally, the focus of the EHP was on environmental health, with a particular emphasis on water, sanitation, and the environment. The model developed for that work was used in conjunction with CARE on the Moyamba Project in Sierra Leone and was successful in reducing water system break-down by 60 percent after the first year (Yacoob et al. 1994). Using the same methods in Belize, Haiti, and Guatemala, NGOs working on water and sanitation were trained to track the effectiveness of health education interventions with specific emphasis on direct practice (Yacoob et al. 1989, 1991, 1992, 1994).

While the model has changed with each use as it has been adapted to local and regional needs, its focus has remained on community participation and establishing relationships between the state and civil society which strengthen local governance and enhance collaboration amongst the different levels of authority and administration and local people. Whether in disease prevention or the reduction of environmental risk, the model weds local needs to political and civil authority through the mechanism of engagement. That engagement may include national level policymakers from health, public works, environment, and local government, such as mayors, creating functional and sustainable partnerships. The model is designed to include individuals from the private sector as well, for instance, from biomedicine, local NGOs, and traditional healers, herbalists, and other practitioners.

All of the applications of the model mentioned in this book focus on health, but it is clear from previous applications that the model is easily adapted to other needs. In Tunisia, the model was employed to create effective partnerships to extend municipal services to underserved peri-urban communities through enhanced interaction and understandings among local policymakers. A series of skill building workshops led to housing improvement, road and bridge construction, and garbage container distribution, and was so successful that the Tunisian government requested and received funding from the World Bank to replicate the process in other towns and provinces (Whiteford 1999). In Benin, West Africa, the model was successfully used to train community members to identify and interrupt disease transmission routes that contributed to elevated childhood diarrhea rates. The project was able to scale up results from similar projects in other countries and facilitate the development of microprojects that included the construction of communal latrines and children's latrines and improvements of the water sources and solid waste landfills. In Bolivia, like Benin, a resource limited country with high levels of child mortality and morbidity due to diarrhea, the model focused on capacity building, community engagement, and the identification of places where families and other authorities could intervene to reduce childhood diarrhea. Success was marked by a reduction by 49 percent of the prevalence of childhood diarrhea in the project communities. The final example we include of how effective and sustainable the model is occurred during the cholera epidemic in rural Ecuador and is the case study provided in more detail later in this book.

The CPI Model

The Community Participatory Involvement (CPI) model has evolved from earlier versions variously referred to as the Community Involvement in the Management of Environmental Pollution (CIMEP) and the Community Participatory Intervention (CPI) models. All share the underlying commitment – as is reflected in the title – that any model of change to be sustained must integrate the community into its design and execution. The five key elements of the CPI model explored in this volume are: 1) the role of effective community engagement; 2) the elicitation and validation of local beliefs; 3) the generation of base-line and follow-up epidemiological data; 4) a scale up with local, regional, and national authorities; and 5) sustained capacity building.

The CPI model emerged from the previous CIMEP work and builds on the engagement of various stakeholders, ranging from the local community to the national government. Central to this engagement is the interlocking structure of the Community Team (CT), the Regional Team (RT), and the Technical Team (TT).

Successful engagement of members of a community requires a sustained commitment of time, a variety of engagement techniques such as knowledge creation workshops, community applications, open community meetings, and change demonstrations. In a diarrhea disease prevention project in Bolivia, for instance, effective community engagement was generated through a technique of household and individual initiatives to understand and prevent childhood diarrhea. While education programs helped generate local and regional commitment, the demonstration of the effectiveness of household water containers (*bidones*) was a visual and public acknowledgment of participation, and became recognized as a reward to engagement. Households could only receive a *bidón* if they completed a variety of activities, such as workshops and community surveys, all of which were designed to enhance and sustain participation (Whiteford 1999).

Figure 3.1. CIMEP Activities Flowchart

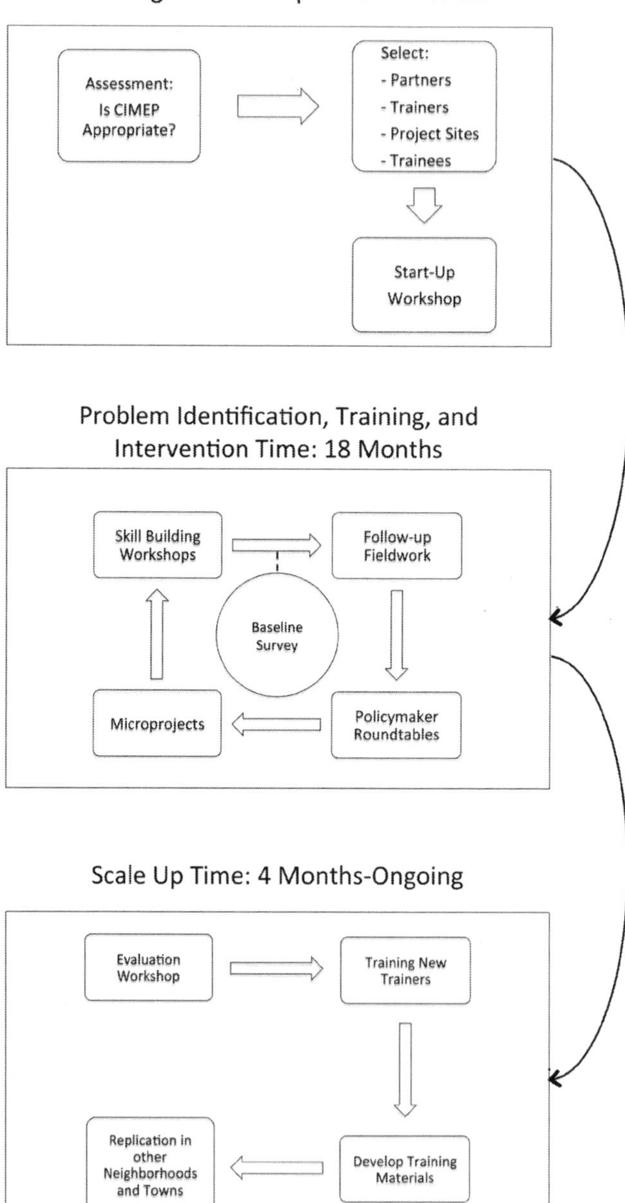

Figure 3.2. CPI Team Model

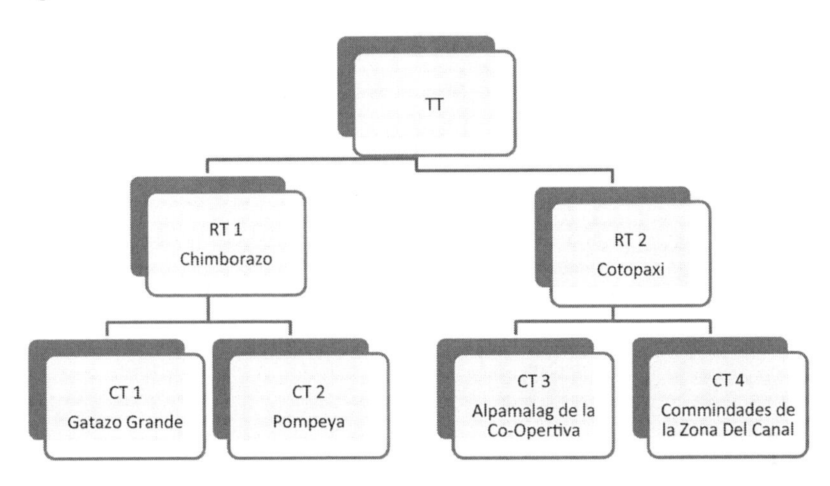

Knowledge generation is multidirectional and is affected by age, power, social status, mode of communication, and rationale or shared need. Too often communities are the targets of information in a unidirectional path, often based on the unacknowledged assumption: "We talk; they listen." As we all know from our personal experiences, there are different learning styles and no one form of communication fits all audiences or all messages. In some situations messages must be translated across the barriers of different languages, histories, ethnicities, and explanatory models. The CPI model bridges these conceptual gulfs by employing a multiplicity of communication styles and by eliciting and validating local knowledge. The model makes real "We learn from you" and "We create knowledge together" types of exchange. We can see this in action, for instance, in the cholera case study we highlight in this book. Non-formal modes of communication were employed during the cholera project workshops, and one of the most effective techniques was to paraphrase what we thought we were hearing the other person/group say until all parties agreed on what they thought was said. This can be a time consuming, and sometimes immensely amusing, process, but in the end it is an effective educational technique.

Acquisition of base-line epidemiological data is also a critical component of the CPI model because only with base-line data can you measure change. Anecdotal accounts are useful because they convey part of the story of how people remember an event, a problem, or a solution. For instance, during the cholera epidemic in Ecuador, a member of one of the study communities told us, "Everyone is dying here and no one comes to help." It was true that many people died, including the speaker's father and husband, and one can honestly imagine how it must have felt to her that everyone was dying and that there was no help. What we learned from her statement was how alone she must have felt, but not the number of people who were infected with cholera or how many died from the disease. In order to measure change, whether to petition the government for help or to document the success or failure of an intervention, both epidemiological information and personal accounts are needed. We need to know how many people actually died each day, if they were clinically diagnosed as having died from cholera or something else, and for how many days people in the community died.

In addition to base-line epidemiological and clinical information, in order to create an effective change plan, we need to define the target population. The CPI model expands the scope of who is defined as a stakeholder from those who live in an affected community to include municipal and regional authorities in the project. To create change in complex settings, a wide variety of players need to be engaged, and roles and responsibilities need to be assessed, clarified, and in many cases, redistributed. For instance, many resource-limited countries, such as the Dominican Republic, face health consequences of vector-borne diseases such as dengue fever. In the Dominican Republic, researchers found that both government authorities and community members understood the causes of the disease and that almost everyone in certain parts of the country had suffered from dengue fever, commonly referred to as 'bone-break fever.' One can imagine that the shared knowledge and experience of the disease would facilitate effective change to reduce its prevalence, but it did not. Instead, what researchers found was almost a standoff between communities and government authorities, each one blaming the other for the lack of action and refusing to make changes until the other one did (Whiteford and Coreil 1997; Whiteford 1997). As a result, even known and effective interventions, such as the removal of standing water, trash collection, or covering outside water containers, did not routinely occur. Each group defined the problem as the responsibility of the other, and until both groups could be engaged as stakeholders, the problem would continue. Community members said the mosquito problem (dengue is carried by the Aedes aegypti mosquito) was the fault of the national government because it failed to have regular trash pickup and removal, and as a result there were standing pools of water everywhere (in banana leaves, old tires, paper trash) in which the mosquito bred. The government, on the other hand, said that the communities were responsible for cleaning up their own neighborhoods and, until they did that, the government was not responsible for collecting trash and hauling it away. This was what the community members called '*una mala unión*' (a bad partnership). The CPI model allows researchers to scale up the results to include decision-makers at the upper levels of the government (clearly necessary in the Dominican Republic example) as well as regional and local level community members.

Theory

The terms 'theory' and 'model' are sometimes confusing, as they are employed differently in distinct disciplines. In this book, we use the term 'theory' to stand for the abstract conceptual connections that work together and reflect what critical elements need to be included in the explanatory framework. For instance, in the CPI cholera intervention, the theoretical framework for the research was of medical ecology. We considered the disease pathogen or agent (the bacteria, *Vibrio cholera*), the mode of transmission (contaminated water), and the host and the host's behavior (fecal-oral routes) all as critically linked components. But in order to understand how people made decisions to change their behaviors to interrupt the transmission of the bacteria, we needed to understand what alternatives were possible for them. In addition, to design effective and sustainable behavior changes, we needed to understand how the people being affected perceived their world and what they saw as realistic alternative options for their behaviors. To do that we enlarged the theoretical structure to include concepts such as culture, history, economics, and politics. Thus, in this book 'theory' refers to an overarching set of linked abstract concepts that identify relations among the component parts necessary to explain the issue under consideration.

On the other hand, the term 'model' is used here to refer to a schematic layout of the central pieces that need to be considered together in order to explain the topic or problem being considered. The traditional biomedical model, for instance, is based on the assumption that "illness can be explained by the presence of pathogens in the body (germ theory of disease) and abnormal physiological processes or biochemical imbalances"

(Coreil 2010:70). Therefore, that model would include information on the pathogen and the resultant disease in the human body. However, the traditional biomedical model leaves little room for the consideration of culturally constructed categories that shape disease access, such as gender, ethnicity, history, and culture.

The CPI model does take into account those culturally constructed categories and is drawn from both a medical anthropology and public health set of environmentally focused theoretical frameworks. The CPI model reflects medical anthropology in an ecological perspective (McElroy and Townsend 2009) and the social ecology of health perspective commonly employed in public health (Coreil 2010). Both theoretical frameworks explicitly embed socio-cultural behaviors in their biomedical and environmental contexts. According to Coreil,

> The [social ecology of health] model is rooted in the traditional public health model of host-agent-environment, which was developed to study the relationship between human populations (host), disease pathogens (agent), and the physical setting (environment) in which people live and work. For instance, the transmission of waterborne illnesses can be studied in terms of the relationship between the human (host) exposed to the disease-causing pathogen (agent) through the use of a contaminated water source (environment). (2010:70)

In our case study of the CPI model in the cholera epidemic, the expanded (to include culture/history/economics) host-agent-environment framework was the cornerstone to understanding the continuing spread of the cholera epidemic in the high Andes.

Central Concepts: Community Participation, Ethnography, Epidemiology, Non-formal Education, and Structural Violence

Community Participation

Anthropological models of behavior-change interventions, like those dependent on community participation, lend themselves to incorporating community members into decision-making and planning, as interventions are designed in collaboration among outside researchers, their local counterparts, and community members (Fals Borda 1990). Community participatory techniques are based on the assumption that communities possess valuable information on their own history, resources, and problems, and they bear the ultimate responsibility for the acceptance and sustainability of any project introduced (Coupal 1995; Cox and Annis 1988). Therefore, the community must be a central part of the process of designing and implementing any change, and through their participation in this change, they acquire a new level of consciousness and engagement that can be used to improve their living conditions (Smith et al. 1997).

The Agua del Pueblo (Water of the People) water and sanitation project from Pacul, Guatemala, provides an example of community-based participatory research (Cox and Annis 1988). This example shows how the incorporation of the community in project design can take it in unexpected directions. The original impetus for the project came when people in the community of Pacul recognized that they needed water and that they could not obtain it without organizing themselves and securing extra-community assistance. They appointed a delegation to talk to an engineer from Agua del Pueblo (a local non-governmental organization). The engineer visited the site, spoke with local people, and made measurements with the community members in order to assess if the project was technically possible. After agreeing to participate, the community members and the engineer signed an agreement that specified the reciprocal obligations of both parties. The community participated in all stages of the project, and that level of engagement was overseen and sustained by a committee of local people elected by the community members to represent them (Cox and Annis 1988).

The committee was in charge of collecting data in the form of a census and of mapping data, fundamental steps in the design process of the latrines and a critical component in the successful implementation of the project. The local committee collected money from the head of each household in the community to build the water system and pay for its maintenance. The local committee was also in charge of an education component in the water project.

In the Pacul water and sanitation project, education became an essential tool in transforming how people in the community disposed of human waste. Pit latrines were the preferred method of waste disposal, and the local committee showed films on sanitation, taught members of the community how to dig latrines, distributed construction materials, helped families with installation, and gave talks on personal and domestic hygiene. The sustained involvement of the local committee increased the likelihood of the success of the project and its replication in nearby communities because the local committee was always present, acting as a local project supervisor and constantly reminding people in Pacul about the importance of clean water and sanitation (Cox and Annis 1988). The local committee was composed of committed supporters of the water and sanitation program and they were known and respected members of the local community. The Pacul project was successful and sustained because of the early and extensive participatory role the community played and how the project empowered the community to stay involved.

One of the lessons learned from this example of effective community-based research is what positive consequences occur when community voices and ideas are respected and validated in the applied research project. While the Water for People project was initially designed as a water supply intervention, during the process of the project, the community requested that latrines be incorporated into the project. And the participatory design was flexible enough to make the change.

Ethnography

Central to medical anthropology is the ethnographic technique of listening to people, capturing what they say and do, and trying to understand that information as a product of its time and place (Schensul and LeCompte 2010). Ethnography requires time for people to come to trust you and to speak meaningfully about sensitive topics. Mariana, the Ecuadorian woman who participated as a member of the Community Team in the CPI cholera project, learned to trust the CPI team and then to share her stories, fears, beliefs, and ideas with us. That, in turn, allowed the team to acquire critical information about local beliefs and practices. While the ethnographic method of asking questions and valuing the answers helped the team develop rapport with members of the community, to be most effective the process must be a process of mutual discovery. As researchers learn about members of the community, community members also learn about the aims, goals and methods of the research project, and the personalities of the researchers. Pivotal to good ethnography is not only the time to acquire trust, but also reasons to legitimate community members' willingness to trust the process and become engaged in it. In Mariana's case, she watched members of her family die during the cholera epidemic; she knew that others in her community and in the communities around hers were dying. She had tried what she knew to prevent their deaths, and she had a great need to trust us and to believe that we could help stop the dying. But the transformation of need into the trust necessary for sustained involvement took working together.

The CPI model draws on the medical ecology framework and on the power of ethnographic research combined with epidemiological mapping (Singer and Clair 2003; Trostle 2005) to produce a picture of local lives set within a larger frame of global distribution of disease. Ethnographic research validates the local frame, collecting insights, perspectives, and personal histories that are embedded in their cultural contexts. It is a technique that allows us to collect and critically analyze the experiences of lives we have not lived and to empathically and intellectually explore their meanings. For instance, when we first met Mariana, she had no reason to believe anything we said, especially since the information was all new and foreign to her beliefs. But in time and through her own need and ability, she told us her experience and how cholera had killed her father, her aunt, and many others – including children – in her community, and of her desperate desire to protect her children from a similar death. From the epidemiological information we knew that Mariana's community had suffered many losses from this disease, and from the ethnographic fieldwork we knew that members of this community, like Mariana, might be willing to engage in a project to control the spread of cholera.

Epidemiology

Epidemiologists, unlike ethnographers, don't often stay in a community long enough to establish trust with the residents. In fact, many epidemiologists are not in the community at all because the discipline of epidemiology focuses on the determinants and distribution of disease and relies on population-based data, as opposed to ethnography that relies on individual stories and experiences. For epidemiologists, what can be measured becomes translated into data, policy, and often practice. Yet, their studies usually do not examine the social processes that are often hidden or obfuscated by supposed 'normality'; rather, they study the numerical patterns and distributions across populations. As a result, those social processes that are shaped, for instance, by rules of gender or power are not visible to nor measured by most epidemiologists. And yet, these 'invisible' patterns of social relations shape the distribution of disease. The CPI model works within the methodological framework of ethnography and relies heavily on understanding both the micro-level of individuals and their communities (and how gender and power shape those experiences) and on how macro-level institutional structures (such as regional or global events) shape policies that become translated into programs and practices. Combined with the power of epidemiological analysis, the ethnography/

epidemiology synergy becomes a powerful tool to understand the determinants of disease.

Non-formal Education

The non-formal education component of the CPI model is a crucial link in the transformation of information into knowledge by way of experience. The model employs a series of workshops (detailed in Chapter 5) to create this transformation. We based our workshops on what Paolo Freire (1970) referred to as 'non-formal education' techniques in which the status differential between teacher and student is removed, and knowledge held by all members of the group is equally respected. In the workshops (and throughout the CPI project), we tried to practice what Freire was referring to when he wrote:

> Education either functions as an instrument which is used to facilitate integration of the younger generation into the logic of the present system and bring about conformity or it becomes the practice of freedom, the means by which men and women deal critically and creatively with reality and discover how to participate in the transformation of their world. (1970:54)

The CPI workshops explicitly tried to mimic "the practice of freedom" that Freire suggested. Theoretically, through education people can become aware of the injustices and inequalities that are present in their everyday lives, and they find tools to challenge those inequities, or they can be blind to those inequities and continue them. The CPI model is framed by this shared value of the importance of equality and the possibility of transformation, and the workshops provided skill sets to build leadership capacity in the communities and facilitate transformations.

The non-formal educational format served as a critical unifier among the various stakeholders in the CPI project. Each project workshop had participation from people living in the local villages as members of the four Community Teams, and also others representing the Ecuadorian Ministry of Health and Well-being, Transportation, Education, some international and local NGOs, and finally there was the Technical Team with representation from the United States Agency for International Development and the Ministry of Health in Quito, the capital city. In short, there were many distinct, and often conflicting, stakeholder points of view represented at the table and in the workshops. What could have been a babel of differences became complicated and complex on-going discussions as each learned what the others were trying to say. The centrality of non-formal education in the CPI model allowed the participants to negotiate among various languages, regions, ethnicities, disciplines, and genders, and among various forms of education by recognizing the importance of these differences, while holding steady the idea of equality. UNESCO (1997:41) defines non-formal education as: "any organized and sustained educational activities that do not correspond exactly to the definition of formal education." While carefully organized with serious attention to detail and content, the workshops fit the UNESCO (1997) definition and were never the same as traditional classrooms. Instead, the workshops were designed to encourage and facilitate integrated interactions among the participants, regardless of their background, training, ethnicity, or gender. As a result, the differences became strengths.

Effective non-formal education sounds as though it is easy to do, but it is not. It requires thinking outside the traditional educational format box, and patience, organization, and open-mindedness. In the cholera project, we were fortunate to hire Dr. Rosario Menendez to facilitate the three workshops and to be a member of the Technical Team. She was an Ecuadorian psychologist trained in non-formal educational techniques, and she brought to the project many years of experience working in communities with non-governmental organizations like Plan International, and with governmental organizations like USAID. Rosario's job was to create conceptually and physically open spaces for the exchange and translation of information. While all participants were required to have some fluency in Spanish, as opposed to the indigenous language which was common among community members, it was not only linguistic translation that was often necessary. It was the worldview exchanges that were most crucial to the project as biomedical approaches, which were largely outside of the indigenous worldview, were combined with more traditional indigenous ideas about disease causality.

While understanding and valuing distinct models of beliefs (for instance, those of biomedically trained health workers and local belief systems of the community members) is a central tenet of the CPI model (e.g., Baer 2011; Kleinman and Benson 2006; Taylor 2003), the non-formal education workshops provided on-the-ground evidence for the importance of negotiating the differences in beliefs to find common ground. For instance, to some community members, geographic places, like a particular isolated river crossing, were believed to be dangerous to one's health, and in the community view, implicated in an explanation of how cholera was transmitted. To elicit community members' explanations of high risk areas, Rosario invited them to take sticks and draw maps in the dirt of the places in the community where one could be exposed to cholera; these maps then provided a template to stimulate local community discussion about how the disease might be transmitted and what those routes of disease transmission might be.

The stick and dirt maps were then transferred to poster-sized paper and posted in a meeting room. And then, with an offer of food and drinks, community members joined the teams in examining and explaining the maps. In the process, those of us not from the communities learned much more than the physical layout of the community. We learned about the historical and social events that shaped the everyday life of the people with whom we were working. As one man explained, "That's where old man Salcedo lived and he just threw his trash out the door every day! And then the dogs would come by, and then the rats would finish it up." In addition to learning about this common practice, we discovered unexpected stories as well. The crossing area over the river that people mentioned as a high-risk area for cholera was where some people collected water. And it was also downstream from the clinic where patients with cholera were being treated and into which clinic staff occasionally dumped patient waste. Suddenly, the invisible connection between local knowledge and biomedical information was made clear as we realized that if contaminated water splashed on someone's hands and they then handled food, they could indeed become ill from crossing the river.

The non–formal workshops also provided opportunities for the exchange of information about water-borne and water-washed diseases and their prevention and cure, as well as providing capacity and skill building training. Rosario often began a workshop with the phrase, *"Como dice mi abuela..."* (As my grandmother says...). And wisdom from her grandmother would then frame the activities and would provide context and connection between local beliefs and the world of biomedicine.

While the idea may or may not have come from her grandmother, in the workshops Rosario demonstrated how to unpack the jargon that often stands in the way of effective communication. She taught the 'power of paraphrase' as we found commonality in references for abstract and otherwise not mutually intelligible concepts. Combined with participatory research, the paraphrase technique showed us how to pare down complex ideas into simple, but essential, statements that made it possible to paraphrase something like 'disease transmission routes' into 'how do you get sick?' In addition, the paraphrase technique provided an effective means to bridge cultural differences by finding underlying similarities in ideas.

The paraphrase technique became a cornerstone to communicating across nationalities, disciplines, and worldviews. It provided a mechanism to incorporate the ethnographically rich information we were being provided into effective outreach and to collaboratively design a culturally appropriate intervention. Thus, Rosario's emphasis on being able to transform the unknown into the known through paraphrase worked on multiple levels: it forced us to find ways to talk about ideas that were meaningful in the local context, to remove jargon that forms barriers to communication, to pare down overly complex presentations of ideas, and to find ways to communicate that were meaningful to all of the groups present.

> *"The poor are not only more likely to suffer, they are also less likely to have their suffering noticed"* (Farmer 2003:344).

Structural Violence

The scholarship behind the concept of 'structural violence' comes from the early part of the twentieth century, introduced by Norwegian Johan Galtung to refer to those often unchallenged institutional practices that continue long-standing inequalities. Legally institutionalized racial, ethnic, religious segregation, whether in the United States, South Africa, or other countries is an example of structural violence because segregation privileges the rights of some groups of people over those of others. And because it reflects the status quo, such segregation is often unrecognized and invisible – except to those whom it harms. Paul Farmer pioneered the application of the concept to health in his foundational work on the structural inequalities in disease and suffering in Haiti (Farmer 1994, 1999, 2004). Many other social scientists have also employed an analysis of macro-level political and economic factors in their understanding of the spread of disease, access to medical treatment, and overall health outcomes (Baer and Singer 2009; Page and Singer 2010; Rylko-Bauer and Farmer 2002; Rylko-Bauer et al. 2009; Singer and Erickson 2011).

What makes structural violence so problematic is that in many cases it is both pervasive and normalized to the point of not being noticed (Walter et al. 2004). For example, Susana Nomales, Community Team member in the project, was eighteen, unmarried, and had few opportunities outside of the family and her home. Gender roles and geography limited her access to education, and while Susana was smart and anxious to continue her schooling into college, she had no resources or support available to do that. Instead, Susana lived at home with her mother, father, five younger brothers and sisters, and her elderly (and, according to Susana, grouchy), unmarried aunt. And if Susana didn't marry soon, she was told, she, too, would become the same as her aunt. The family expectations for Susana were that she would either marry and leave the household to move into her husband's family house, or she would stay and continue to care for her younger brothers and sisters, her parents, and her elderly and grouchy aunt. Cultural expectations about power and gender constrained and limited Susana's ability to attain the education she desired, which in turn shaped her employment opportunities and her access to extra-familial social networks that could be used for upward mobility. Geography and ethnicity also limited her access to education because of the lack of secondary schools in her community and her family's reluctance to allow her to go to another, larger community, where she could attend high school. Susana felt trapped by the limitations placed by her gender role and, while unsure how to change anything, was dissatisfied with how reduced her own agency was. She wanted to make changes in her own life, but until cholera and the CPI project came to her community and brought outside ideas about how to control the spread of cholera and with it, ideas about social roles and access to health care and education, Susana saw few options for herself. But she learned to see them.

When gender limitations are taken for granted and normalized, they become part of an unquestioned routine that privileges the rights and roles of some over those of others. As Susana's case shows us, gender roles in her community gave power to men to acquire resources such as education and denied that same opportunity to women, especially to unmarried, young women. What Scheper-Hughes (1992) and others have referred to as "everyday violence" is violence that occurs so commonly that it is neither questioned nor noticed, yet daily constrains the life possibilities of some while favoring others. In the CPI model, the incorporation of the concept of structural and everyday violence reminds us to look at the social and institutional practices that reduce equality, without which incorporation we cannot understand the larger social context.

> *"Structural violence is silent, it does not show – it is essentially static; it is the tranquil waters"* (Galtung 1969:173).

The concept of 'structural violence' has implications for public health policy and programs. For example, the lack of health posts or potable water in rural, indigenous areas can be seen as the result of a history of privileging non-native people living in urban centers. The concept of structural violence directs our attention at the micro-level to institutions and their effect on people, and simultaneously, at the macro-level to the historical and economic forces that shape those institutions. In the Ecuadorian CPI project, we used the ideas behind structural violence to identify the cultural and historical dimensions of the epidemic and how structural violence shaped access to care, and we relied on the idea of everyday violence to show how and why the disease shaped the lives of its sufferers. Similarly, Walter, Bourgois, and Loinaz (2004) applied the idea of structural violence to gendered experiences reported on in the public health literature about illness and disability to uncover and identify their historical and cultural roots.

The concept of 'everyday violence' is useful on multiple levels. It provides an opportunity to identify how lived experiences, or the "micro-level experiences," create heightened risk for disease and suffering in addition to the historical and economic forces identified as structural violence. In this way, the macro-level forces are not dissociated from micro-level experiences of everyday life that are central to diseases such as cholera, where individualized experiences can foster increased risk for diseases.

The combined structural and everyday forms of violence become embodied in, for instance, Mariana's experience. "We didn't know about germs. We didn't know to wash our hands after going to the bathroom and when we cooked." This comment highlights her previous lack of understanding of biomedical explanations of fecal-oral disease routes of transmission (the daily violence of rural poverty), but also identifies the structural violence of her environment where she had no indoor sanitation or potable water, and, therefore, an increased risk of disease and death. While the lack of access to safe water and sewage systems and the scarcity of health posts seemed normal to Mariana and Susana, they were both aware that the cholera epidemic lasted longer in their communities than it did in other communities that had clean water and sanitation.

In this chapter we have presented the CPI model and its development and theoretical basis. The concepts and methodologies derived from community participation, ethnography, epidemiology, non-formal education, and, finally, structural and everyday violence that are central to the CPI model are introduced and briefly explained. In the next chapter we present the epidemiology of the disease of cholera both globally and in Ecuador, where the CPI cholera project took place.

Chapter Summary

- The CPI model was developed and funded by the United States Agency for International Development (USAID) and was implemented by the Environmental Health Project (EHP). It has five key elements: 1) the role of effective community engagement; 2) elicitation and validation of local beliefs; 3) generation of base-line and follow-up epidemiological data; 4) scale up with local, regional, national authorities; and 5) sustained capacity building.
- The CPI model rests on the engagement of various stakeholders, ranging from the local community to the national government. It expands the scope of who is defined as a stakeholder from just those who live in an affected community by explicitly identifying both local community members and municipal and regional authorities as stakeholders in the project. For example, the CPI model used in Ecuador centered on the interlocking structure of the Community Team (CT), the Regional Team (RT), and the Technical Team (TT), all of whom played an important role in the design, implementation, and sustainability of the changes in the community. This technique of interlocking horizontal and vertical teams epitomizes what is often referred to as diagonal integration and facilitates replicability and sustainability of the methods used.
- In the CPI cholera intervention, the theoretical framework for the research was of medical ecology, where the disease pathogen or agent (the cholera vibrio bacteria), the mode of transmission (contaminated water), and the host and the host's behavior (fecal-oral routes) were considered as critically linked components.
- Central concepts of the CPI model:
    - *Community participation*: Community participatory techniques are based on the assumption that communities possess valuable information on their own history, resources, and problems and that they bear the ultimate responsibility for the acceptance and sustainability of any project introduced. Therefore, the community must be a central part of the process of designing and implementing any change.

- *Ethnography*: While the ethnographic method of asking questions and valuing the answers helped the team develop rapport with members of the community, to be most effective it must be a process of mutual discovery. As researchers learn about members of the community, community members also learn about the aims, goals, and methods of the research project and the personalities of the researchers. Pivotal to good ethnography is not only the time to acquire trust, but also reasons to legitimate community members' willingness to trust the process and become engaged in it.
- *Epidemiology*: the discipline of epidemiology focuses on the determinants and distribution of disease and relies on population-based data. Acquisition of base-line epidemiological data is a critical component of the CPI model because only by collecting base-line data can you measure change.
- *Non-formal education*: The CPI model is based on the idea that through education people can become aware of the injustices and inequalities that are present in their everyday lives, and they find tools to challenge those inequities, or they can be blind to those inequities and continue them. The CPI model is framed by the shared values of equality and the possibility of transformation, and the workshops provided skill sets to build leadership capacity in the communities and facilitate transformations.
- *Structural violence*: The concept of structural violence directs our attention to the micro-level of institutions and their effect on people and, simultaneously, to the macro-level of the historical and economic forces that shape those institutions. The CPI model in Ecuador used the ideas behind structural violence to identify the cultural and historical dimensions of the epidemic and how it shaped access to care and relied on the idea of everyday violence to show how and why the disease shaped the lives of its sufferers.

In-class Exercise: Community Perception Maps

Activity: Gather information about people's perceptions about places and other kinds of information. It is important to differentiate between a street map, for instance, and a map based on memories. In this case, we want to find out how people translate their perceptions of both kinds of maps into drawings.

Goal: Be able to compare various participants' perceptions of a common place.

Method: Hand out paper and pencils and ask participants to draw, for instance, the location of the building they are in and the immediately surrounding landscape. Ask them to include any locations that may be dangerous to someone not familiar with the building or the area.

1. Compare the drawings and analyze the differences, and ask students to explain why they included what they did.
2. Summarize what kinds of different perceptions were articulated by different people in the group and what those differences tell about how danger or risk are perceived?

Observe the variables and heterogeneity in the group being studied and how they describe their perceptions of the world around them. Include what personal categories might shape individual perceptions (gender, age, nationality).

Chapter Four

The Disease

The Cholera Epidemic in Ecuador

Cholera, an acute bacterial infection of the intestine, is caused by the bacterium *Vibrio cholerae* (see Figure 4.1) (WHO 2003). Cholera is acquired through consuming contaminated food or water or contact with contaminated feces (fecal-oral route) (WHO 2003). Its main symptoms include watery diarrhea and vomiting. If left untreated, these can lead to severe dehydration and result in death (WHO 2012).

According to the World Health Organization (WHO) about 75 percent of people infected with the cholera vibrio do not develop any symptoms, although the bacteria are present in their feces for 7–14 days after infection and are shed back into the environment, potentially infecting other people. Among those people who do develop symptoms, 80 percent have mild or moderate symptoms, but close to 20 percent develop acute watery diarrhea with severe dehydration that can lead to death if untreated. People at greatest risk of dying from cholera are those with low immunity – such as malnourished children, the elderly, or people living with certain chronic diseases (WHO 2012). Areas at risk for promoting the spread of cholera are those without proper water and sanitation infrastructures (WHO 2012).

Community Participatory Involvement: A Sustainable Model for Global Public Health
by Linda M. Whiteford and Cecilia Vindrola-Padros, 71–85.

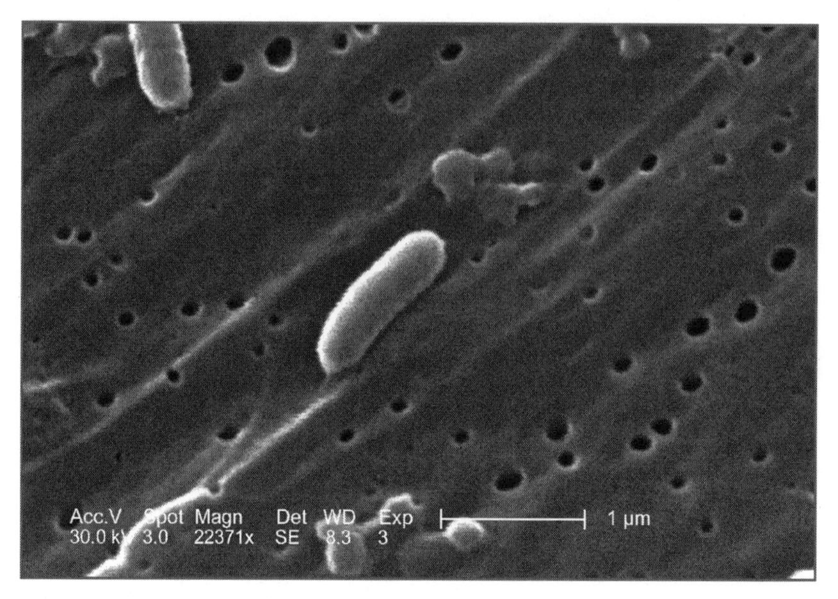

Figure 4.1. Electron micrograph (SEM) of the bacterium *Vibrio cholera*, Sero-group 01

A Brief History of Cholera

There have been eight documented cholera pandemics. A pandemic is an epidemic that occurs worldwide or over a wide geographical area, often affecting multiple countries and even continents. The timeline in Figure 4.2 provides an overview of the eight cholera pandemics.

The first documented cholera pandemic, 1817–1823, spread from India to Southeast Asia, Central Asia, and the Middle East. The main reason for the spread of the disease was the increase in human interaction mainly produced by new ways and causes of travel (immigration, military campaigns, trade) (Lee and Dodgson 2000).

The second pandemic of 1826 to 1838 lasted longer and extended to new areas in Europe and the Americas (Canada, United States, Latin America, and the Caribbean). The large influx of European migrants to North America during this period led to the introduction of the *Vibrio cholerae* in New York in June 1832 (Lee and Dodgson 2000:218). The intercontinental railway facilitated its spread across the country and further south to the Caribbean and South America (Speck 1993).

The third pandemic spanned over sixteen years (1839 to 1855) and spread from India to Afghanistan with British troops (Speck 1993).

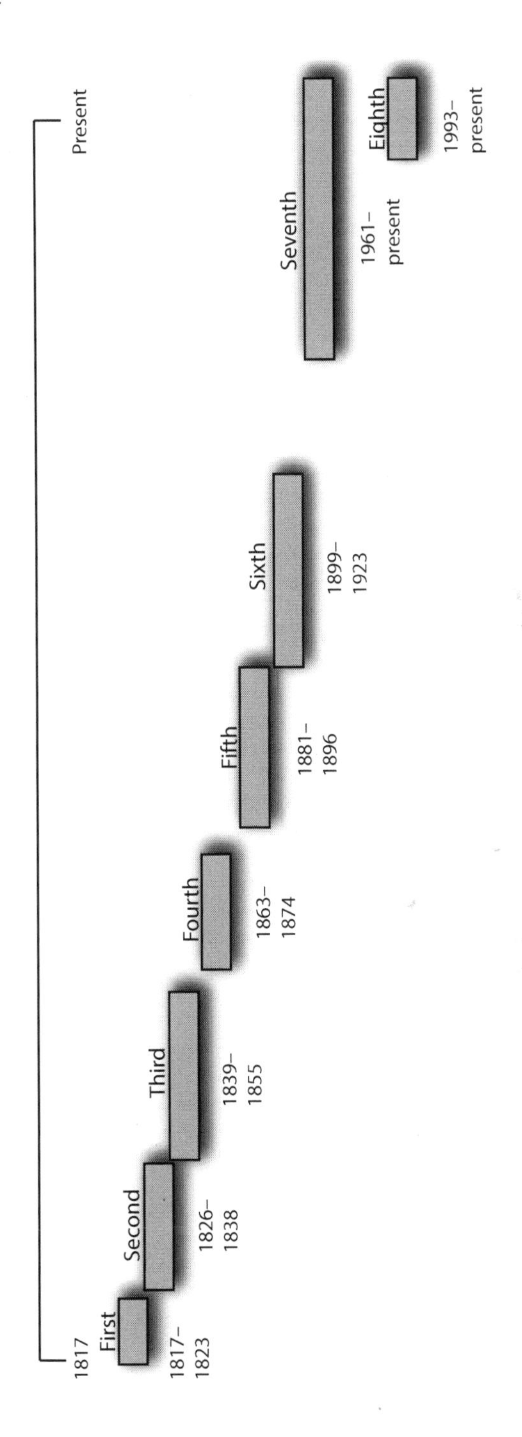

Figure 4.2 Cholera pandemics, 1817–present

Similar to the case of the second pandemic, the disease spread across different continents, affecting North Africa, Europe, and the Americas. It was during this pandemic that John Snow mapped the large number of deaths occurring in London in the late 1840s and interviewed surviving family members (Johnson 2006). The information he was able to collect led him to believe that the disease was water-borne and that its spread resulted from the confluence of several separate and distinct private water systems distributed through the Broad Street water pump (Johnson 2006).

The fourth pandemic (1863 to 1874) followed a similar trend and was exacerbated by the opening of the Panama Canal. The fifth pandemic took place from 1881 to 1896, and it was in 1883 that Robert Koch identified the bacillus responsible for the disease through studies he performed in Alexandria and Calcutta (Lee and Dodgson 2000; Speck 1993). The sixth pandemic (1899 to 1923) endured a spread of the disease over an unprecedented period of time, 24 years. It affected Asia, Africa, and Europe, but it failed to reach the Americas (Speck 1993).

The seventh pandemic that began in Indonesia in 1961 was probably due to civil conflict, which led to migration and military intervention (Barua 1972). Cholera then spread across Southeast Asia and the Western Pacific, then stretched to the Middle East, and ultimately made its way to Europe and the Americas. This last pandemic before the 1990s is usually divided in three stages: 1961–1969, 1970–1977, and 1978–1989, with the first two stages representing the peak (Lee and Dodgson 2000:222). A total of 1.72 million cases of cholera from 117 countries were reported during the period prior to the 1990s (Narkevich et al. 1993). In the early 1990s, cholera appeared and spread across Latin America, producing 1.4 million cases and 10,000 deaths in nineteen countries (Sanchez and Taylor 1997). This was the longest pandemic and the one with the widest geographical spread (Lee and Dodgson 2000).

In 1992 a new derivative of the El Tor strain of cholera emerged in Bangladesh and resulted in a widespread epidemic within the country (WHO 2006). This new sero-group was named "*V. cholerae* 0139 Bengal" and was reported to exist in at least a dozen countries. The need for close surveillance of the new strains of cholera was noted, and WHO cautioned about the possibility of a pandemic, which further emphasized the need for careful observation (WHO 2006). The emergence of this new sero-group led to the eighth cholera pandemic, which spread from Bangladesh to Pakistan, Thailand, ten other Southeast Asian countries, and the United States (Lee and Dodgson 2000:222).

Cholera in the Twenty-first Century

The seventh and eighth pandemics are considered active across the world through the first quarter of the twenty-first century. A series of recent outbreaks has influenced the way we currently think about and deal with cholera epidemics. In April 2002, there was a resurgence of cholera in Bangladesh. Approximately 2,350 people with *V. cholerae* 0139 Bengal were admitted to the Dhaka Hospital of the International Centre of Diarrheal Disease Research within a two month period at the beginning of 2002. Preliminary reports estimated that more than 30,000 cases of cholera occurred in Dhaka and the surrounding areas (Faruque 2003:1116). Studies monitoring these outbreaks indicate that strains of the 0139 sero-group were undergoing rapid genetic changes and on-going research provided new insights into the epidemiology of the disease (Faruque 2003:1116).

Another outbreak of cholera – caused by food-borne transmission as a result of contaminated raw vegetables – occurred in Lusaka, Zambia, in 2003–2004 (MMWR 2004:785). Cholera had become almost commonplace in Zambia, which had cholera epidemics in 1991, 1992, and 1999. In 1998, the Safe Water System (SWS), a point-of-use water disinfection and storage strategy, was launched by a partnership of organizations, including USAID and CDC. Unfortunately, while there was documented widespread acceptance of the SWS in Lusaka's cholera-affected communities, the investigation in the 2004 outbreak suggested that fewer than 20 percent of Lusaka's shantytown residents purchased the Clorin® cleaning solution to add to their water (MMWR 2004:785). So even with a protected water supply system, the failure to be able to afford the disinfecting solution rendered the water system open for contamination.

In January 2005, Médecins Sans Frontières (MSF), or Doctors Without Borders, re-opened a treatment center in Bujumbura, Burundi, to treat a new outbreak of cholera. During the first three days of the outbreak, before the government declared an epidemic and provided free health care, people died because they could not afford to pay the five dollar hospital entrance fee (MSF 2005).

In Southern Sudan, the world's newest country, it was reported that an outbreak in early 2006 resulted in a total number of cases had reached nearly 9,000 with 238 deaths (IRIN 2006). In response, WHO recommended control measures, such as strengthening surveillance systems, standardizing case management, implementing environmental control

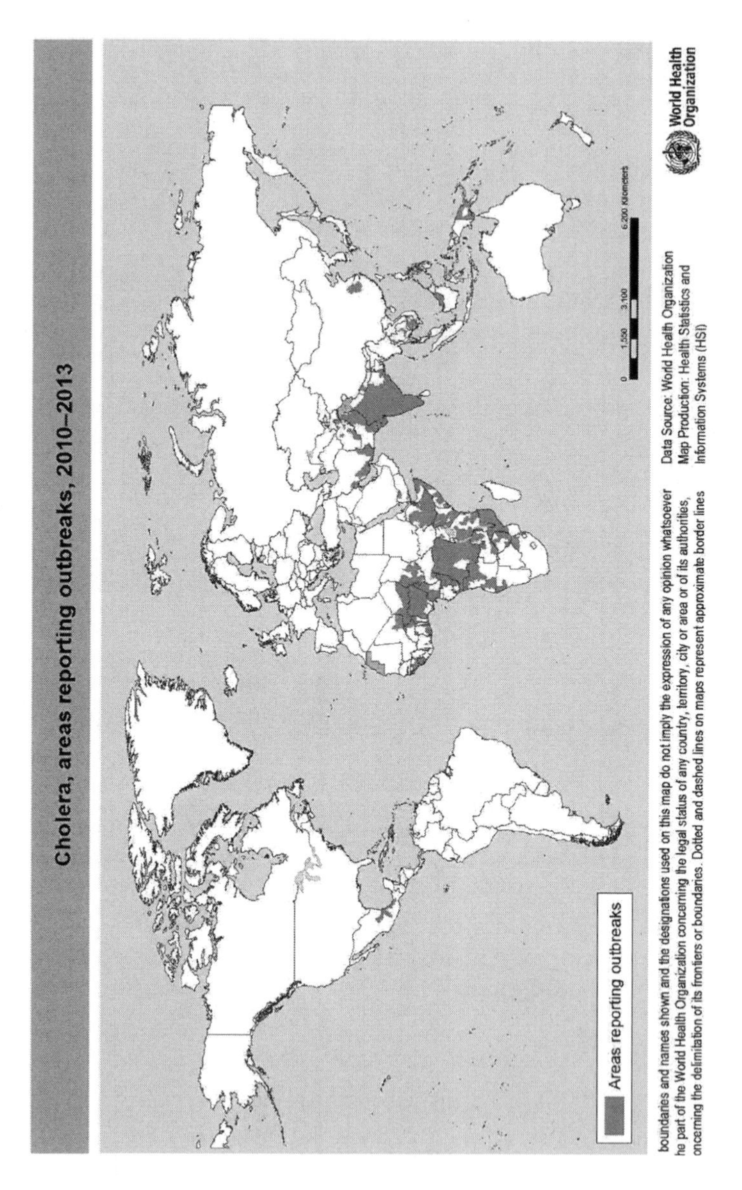

Figure 4.3. Global distribution of cholera cases, 2010–2013, WHO

measures such as the chlorination of public water supplies, and conducting health education and hygiene promotion campaigns (WHO 2006).

A review of epidemiological records produced by the WHO for 2012 show that cholera continues to represent a worldwide burden, with an

Figure 4.4. Countries reporting cholera deaths in 2012, WHO

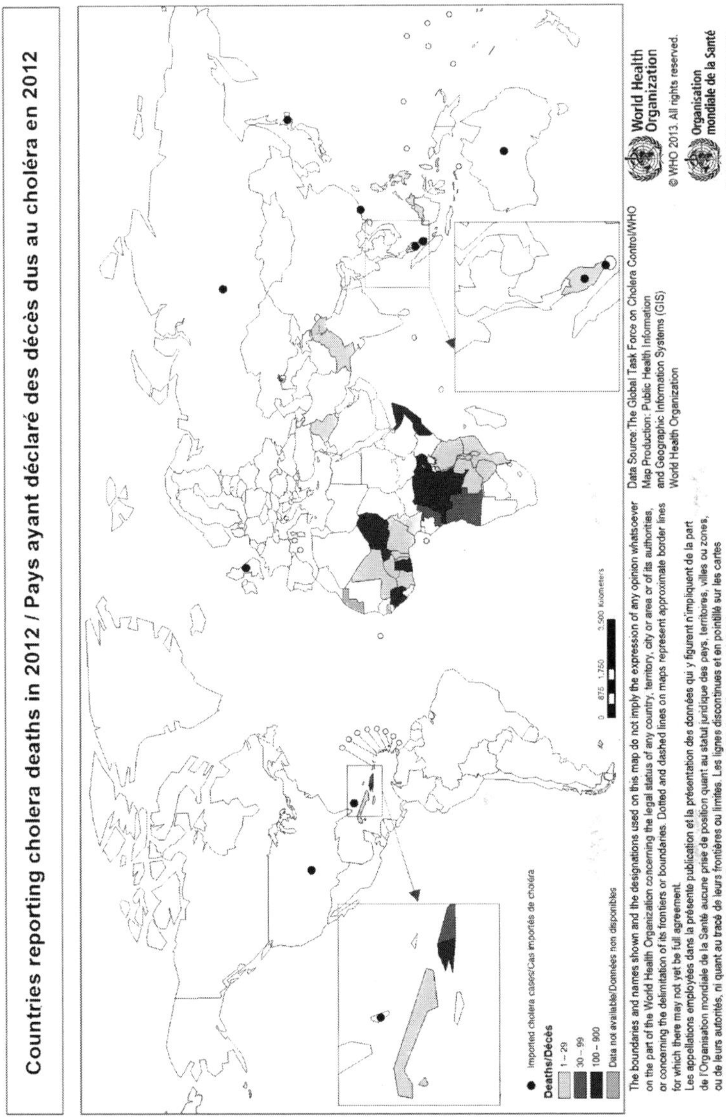

estimated 3–5 million cases and over 100,000 deaths each year (CDC 2013; WHO 2012, 2013). As Figure 4.3 shows, even though the number of cholera cases reported in 2012 decreased when compared to previous years, a total of 48 countries from all continents reported cases and deaths (WHO 2013).

As Figure 4.4 indicates, 23 countries from Africa accounted for 67 percent of the deaths worldwide, and the Americas reported 31 percent of the global total of deaths (WHO 2013). The high rate from the Americas was mainly produced by the cholera epidemic affecting Haiti and the Dominican Republic after the earthquake that devastated Haiti in 2010 (WHO 2013). By 2011, 470,000 cases and 6,631 deaths had been reported (CDC 2011). The numbers of cases and deaths produced by cholera have decreased since the epidemic experienced its peak in 2011, but the outbreak continues to be a reason for concern in the region. The CDC has labeled this outbreak as "the worst cholera outbreak in recent history" (CDC 2011).

The WHO has responded to these recent cholera outbreaks by recognizing the re-emergence of this disease as a "significant public health burden" in resolution WHA 64.15 (WHO 2013:326). As forms of control, the WHO has recommended the use of antibiotic agents in cases of moderate and severe dehydration, treatment with zinc, and the expansion of the use of a two-dose oral cholera vaccine (Waldman et al. 2013). In regards to prevention, much work still needs to be done to guarantee safe drinking water and the adequate disposal of waste (Waldman et al. 2013).

Cholera in Latin America

Cholera had disappeared from Latin America for almost 100 years until it appeared in Peru in January 1991 (Gabastou et al. 2002). The disease spread rapidly, reaching Ecuador on February 20 of the same year (Barragan Arenas 1991:34). On March 1, nine cases of watery diarrhea, vomit and dehydration were reported in Hospital de Machala, in the Ecuadorian province of El Oro, and after two months, 13,904 cases had been confirmed and 232 people had died (Barragan Arenas 1991:34). During the first three years, cholera reached all countries in Latin America, with the exception of Uruguay and the Caribbean Islands (see Table 4.1) (Bahamonde Harvez and Stuardo Avila 2013:41). Colombia felt a dramatic increase in cases in 1991, while Brazil and Bolivia saw a substantial increase in 1992 (Chevallier et al. 2004).

During the mid-1990s the number of reported cases started to decline, but a recurrence was seen in 1998, due to the El Niño phenomenon (mainly felt in Ecuador and Peru) and the damage to the sanitation system produced by Hurricane Mitch in Guatemala, Honduras, and Nicaragua (Bahamonde Harvez and Stuardo Avila 2013:41; Gabastou et al. 2002).

Table 4.1. Cholera Cases Reported to WHO (Selected Countries), 1991–1998

Country	1991	1992	1993	1994	1995	1996	1997	1998
Argentina		553	2,080	889	188	474	637	12
Belize		159	135	6	19	26	2	28
Bolivia	206	22,260	10,134	2,710	2,293	2,847	1,632	466
Brazil	1,567	30,309	59,212	49,455	15,915	4,634	2,881	2,571
Chile	41	73	32	1		1	4	24
Colombia	11,979	15,129	230	996	1,922	4,428	1,508	442
Costa Rica		12	14	38	24	19	1	
Ecuador	46,320	31,870	6,833	1,785	2,160	1,059	65	3,724
El Salvador	947	8,106	6,573	11,739	2,923	182		8
Guatemala	3,674	15,395	30,604	5,282	7,970	1,568	1,263	5,970
Honduras	11	384	4,007	4,965	4,717	708	90	306
Mexico	2,690	8,162	10,712	4,059	16,430	1,088	2,356	71
Nicaragua	1	3,067	6,631	7,821	8,825	2,813	1,283	1,437
Panama	1,178	2,416	42					
Peru	322,562	212,642	71,448	23,887	22,397	4,518	3,483	41,717
Venezuela	15	2,842	409			269	2,551	313

(Source: WHO Global Health Observatory Data Repository 2015)

Spread of Cholera in Ecuador

Research carried out by PAHO (Pan American Health Organization) pointed to Campamento La Puntilla in the El Oro Province, Ecuador, as the initial source of contamination. Cholera reached Ecuador in 1991 when a Peruvian fisherman traveled from Tumbez (in Peru) to the Ecuadorian coast for the collection of shrimp larvae (Malavade et al. 2011). Two weeks prior to the first reported case in Ecuador, cholera cases had already been reported in Tumbez. The *Vibrio cholerae* O:1, "El Tor," was found in a nearby water source after a septic tank overflowed, with contaminated fecal water reaching the drinking water.

The epidemic in Ecuador started on the Pacific coast where the Peruvian fisherman collected shrimp larvae (week one), then disseminated to El Oro (week nine) and the provinces in the Coastal region (Gabastou et al. 2002). This was due in part to people escaping the area when cholera cases were announced to the public and to their returning to their places of origin to attend the celebrations of Holy Week (*Semana Santa*). The disease then reached the communities of the Andean highlands (*Sierra*), causing a peak in cases of mortality due to the poor living conditions (lack of potable water and sanitation) of the population and their lack of access to medical attention. The increase in number of cases by region is presented in Table 4.2.

The following factors were identified as contributors to the spread of the disease:

1. Clandestine and low quality water connections to the national supply system did not allow the treatment of water in the same way as the water coming from the main system, resulting in contamination in marginal and low-income areas.
2. Direct contamination of water sources due to open defecation, overflowing septic tanks, or the dumping of hospital sewage.
3. The consumption of contaminated food, such as raw seafood for the preparation of ceviche (a popular dish)
4. The consumption of food from street vendors where the utensils and food were not washed properly.
5. Directly dipping drinking vessels into containers for storing water. (Malavade et al. 2011:3; Weber et al. 1994)

Controlling the Epidemic in Urban Ecuador

In urban Ecuador the cholera epidemic was relatively quickly controlled through a variety of traditional public health techniques, such as a massive health educational campaign to get citizens to wash their hands, to avoid eating food prepared and sold on the street, to avoid raw shellfish or other uncooked sea-food, and to use toilets, rather than empty lots and back alleys, for defecation. But, as we will see in the next chapters, while these techniques were effective in the urban areas, they failed to control the epidemic in the rural, indigenous areas of the high Andes. The cultural practice of consuming traditional meals like ceviche (a meal of uncooked seafood like shrimp, mussels, and fish) increases the possibility of passing on the cholera vibrio. These two routes (failure of sanitary and hygienic

Table 4.2. Number of Cases and Number of Deaths by Region in 1991

Region	Province	Number of cases	Number of deaths
Highlands (Sierra)	Azuay	448	15
	Bolivar	81	7
	Carchi	38	1
	Cañar	706	15
	Chimborazo	3,140	107
	Cotopaxi	2,177	92
	Imbabura	4,745	85
	Loja	321	14
	Pichincha	2,127	34
	Tungurahua	1,732	47
Coast	El Oro	4,673	23
	Esmeraldas	5,425	52
	Guayas	14,951	119
	Los Ríos	3,814	32
	Manabí	1,845	49
Amazonian	Morona	3	3
	Napo	1	0
	Pastaza	34	1
	Sucumbíos	51	1
	Zamora	5	2
Galapagos Islands	Galapagos Islands	3	0

(Source: Dirección Nacional de Epidemiología)

practices of handwashing/feces disposal, and the handling and eating of uncooked foods such as seafish) of cholera dissemination were identified early and targeted in urban centers, such as Quito and Guayaquil, and although street venders and seafood sellers suffered financially from the

Table 4.3. Annual Number of Cases of Cholera and Deaths in Ecuador, 1991–2004

	Number of Cases	Number of Deaths
1991	46,320	697
1992	31,870	208
1993	6,833	72
1994	1,785	16
1995	2,160	23
1996	1,059	12
1997	65	3
1998	3,724	37
1999	90	0
2000	27	1
2001	9	0
2002	0	0
2003	25	0
2004	5	0

(Source: Dirección Nacional de Epidemiología)

results of the public health campaign, it did reduce the spread of cholera in the cities.

A second major thrust of the urban campaign to control cholera was to provide waste removal and public toilets. Previous to the cholera epidemic, in urban centers such as Quito it was close to impossible to find a public toilet. As a result, people had little recourse other than to use unoccupied spaces (back alleys, empty lots, and so on). The interventions were classic public health interventions, such as disease education, provision of safe or uncontaminated water, and proper sanitation, and they were successful in interrupting the transmission of cholera in the urban areas.

Treatment of Cholera Cases

It is estimated that during the first year of the epidemic, 81 percent of cholera cases were treated in hospitals (Creamer et al. 1999). One of the main reasons for this was that intravenous fluid replacement was the therapy of choice (Malavade et al. 2011). This led to the overburdening of hospital facilities, delays in treatment, and an increase in treatment costs (Creamer et al. 1999).

In light of this situation, government authorities began to promote the use of oral rehydration therapy. Oral Rehydration Therapy (ORT), developed in the late 1960s by Dr. Richard Cash and pilot-tested in Bangladesh (Nalin and Cash 1971), is based on the simple combination of water, salt, sugar, and potassium. Oral Rehydration Salts (ORS) do not treat the underlying causes of diarrhea, the severity of its symptoms, or the disease causing the diarrhea. ORS give the infected person a chance to live long enough to combat the disease through other means, such as antibiotics, or to allow the pathogen time to pass through his or her digestive system.

This strategy helped reduce the initial burden placed on hospitals, as ORS were distributed among the general population in small packets and could be consumed at home or in local health clinics.

This chapter has presented the epidemiology of cholera, its world-wide distribution, and the eight global pandemics since the disease was first identified and assessed. It is a disease associated with the failure to provide people with access to a safe and reliable system of potable water and a lack of sanitation. The course of the disease is also shaped by other markers of poverty – poor nutrition and education. And, the spread of the disease is also shaped by socio-cultural variables such as gender roles, power and social status, ethnicity and religion, and most of all, access to affordable health care and education. As we see from the case study of the cholera epidemic in the high Andes of Ecuador presented in the next chapter, and the use of the CPI behavior change model, when the socio-cultural variables as well as the bio-medical ones are taken into account, communities can control the spread of the disease.

Chapter Summary

- Cholera is caused by the bacterium *Vibrio cholera*.
- Cholera is acquired through consuming contaminated food or water or contact with contaminated feces.
- There have been eight documented cholera pandemics.
- Cholera continues to represent a worldwide burden with an estimated 3–5 million cases and over 100,000 deaths each year.
- Cholera re-emerged in Latin America in 1991 and spread rapidly from Peru to Ecuador.
- The main factors contributing to the spread of the epidemic in Ecuador were: the direct contamination of water sources (open defecation or hospital waste disposal), unsuitable mechanisms for treating water, and the consumption of contaminated food (raw seafood and food from vendors).
- Different campaigns promoting the treatment of water and availability of public toilets were implemented in large cities in Ecuador, but the campaigns did not reach many rural areas.

In-class Exercise: Tracking and Controlling Epidemics

Activity: Search for epidemiological information on a recent infectious disease outbreak like the Ebola outbreak in 2014 or the cholera outbreak in Haiti that began in 2010. For example, use the CDC and WHO websites.

Goal: Understand how to use epidemiological information for community-based behavior change.

Method: Use existing epidemiological data to track an epidemic by looking for the following information:

1. Where and when did the epidemic start?
    a. What was its spatial distribution (how did it spread geographically)?
    b. How much time did it take to spread?
    c. What were the main factors contributing to the spread?
    d. Was it controlled, and if so, how?
    e. How were cases treated (medically, by the press, by other authorities)?

Using the same information, write a short essay discussing the lessons government officials, public health practitioners, and healthcare professionals learned while dealing with the epidemic. What could be done to prevent the epidemic from happening again?

Chapter Five

Case Study

The CPI Model in Ecuador – The Cholera Project

Introduction

This chapter provides an overview of the steps that preceded the actual application of the CPI model, a description of the research team skills as required by the model, and the sequence in conducting the research. In addition, the chapter includes specific information on the CPI cholera project workshops and examples of the community-based activities, such as interviews and household observations.

From October 1994 through October 1995 the CPI model was used to slow and then stop the spread of a cholera epidemic in the highlands of Ecuador. As described in Chapter 4, in 1991 the disease spread from the coast of Peru through southwestern Ecuador, through the urban areas and into the high Andes. Now, we provide the story of how the CPI model changed people's lives. The medical ecology perspective was employed to focus on the unit of the household and its use of resources, particularly water. Using a medical ecology framework allowed us to situate the communities within their geophysical reality (high mountains, snow-melt

Community Participatory Involvement: A Sustainable Model for Global Public Health
by Linda M. Whiteford and Cecilia Vindrola-Padros, 87–118.

rivers, dispersed hamlets), in the geopolitical landscape (marginalized, rural, indigenous), and as embedded in the historical and cultural cosmology of being Quechua descendants of the Inca.

The CPI Planning Stage

The CPI Model is based on the following steps: 1) conduct a baseline study, 2) pilot test, adapt, and then conduct the extended education program and behavior change interventions, and 3) conduct follow-up research

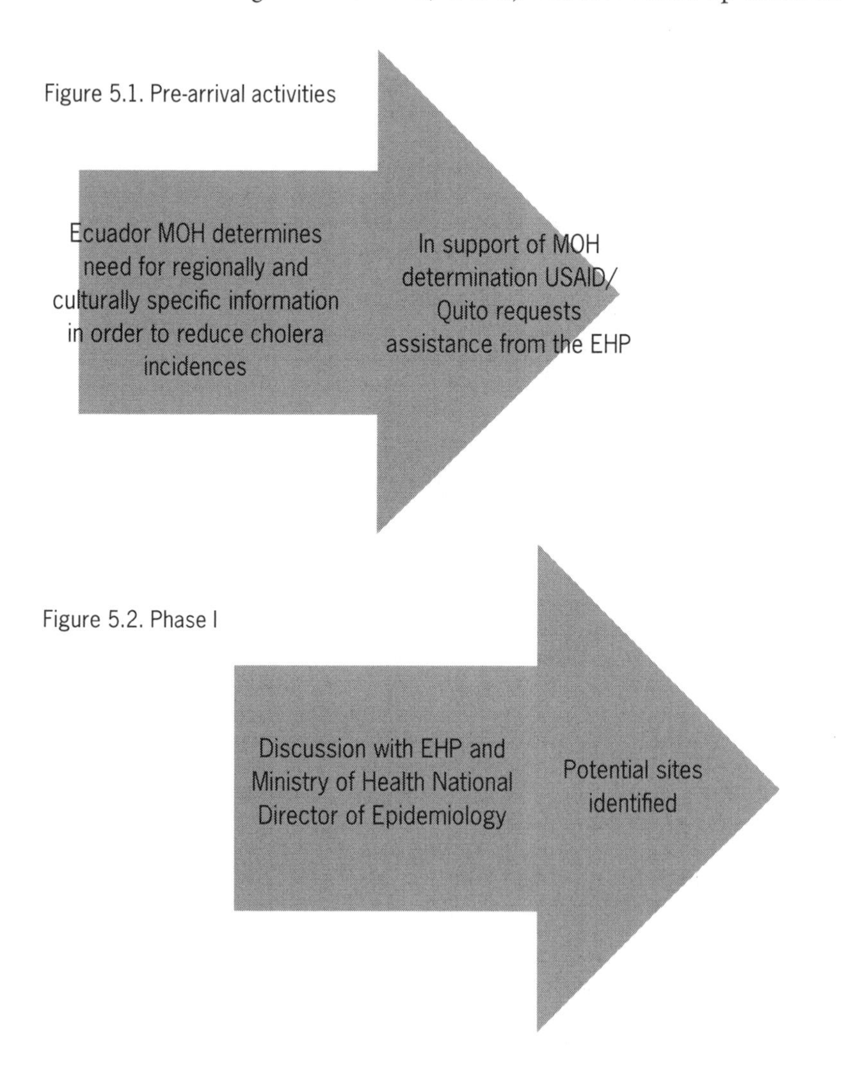

Figure 5.1. Pre-arrival activities

Figure 5.2. Phase I

on risk-related behaviors and an evaluation of the overall activity (see 'Outcomes' in Chapter 6). To accomplish these goals, the project included two national seminars to provide information about the project and the CPI model to high-level policy-makers from the Ecuadorian government, to leaders in the private sectors, and to authorities in private voluntary organizations (PVOs) and NGOs working in Ecuador. Information gathered during these two early seminars was incorporated into the project activities and was used to enhance the follow-up evaluation.

Figure 5.3. Phase II: Technical Team visits potential sites

Team staff visits potential sites & interviews relevant state directors

Interested state directors invited to attend meetings in capital

Epidemiologists prepare meeting materials for MOH staff epidemiologist for social/physical epidemiology of endemic state specific cholera

State directors provide overview of activities, cooperative agreements, interinstitutional working groups, community–based sources

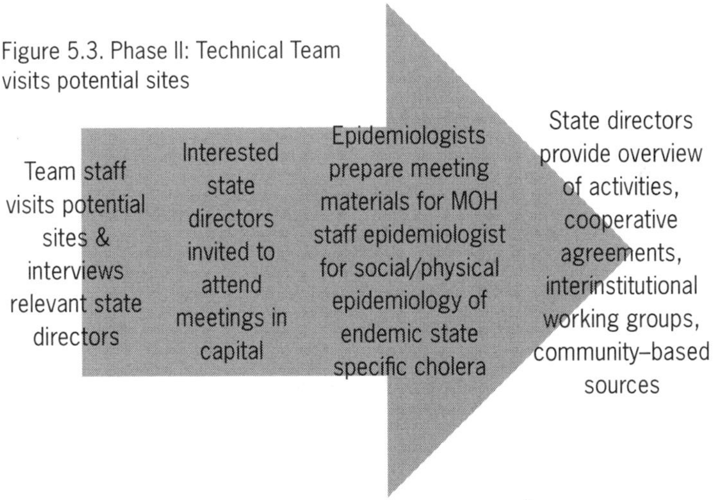

Figure 5.4. Phase III: Quito directors and staff epidemiologist presentation meeting

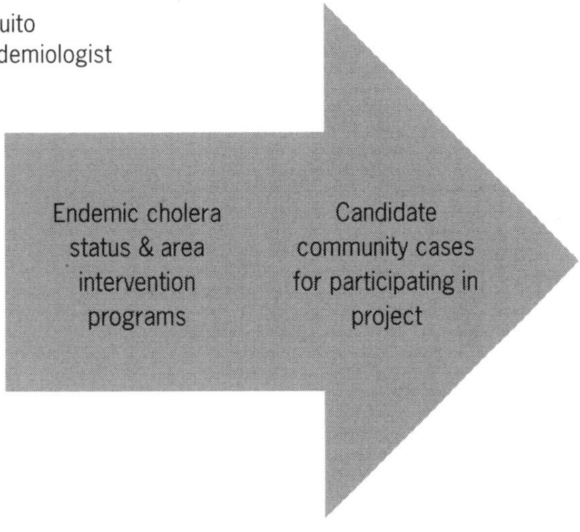

Endemic cholera status & area intervention programs

Candidate community cases for participating in project

In addition to the national-level presentations to government officials and policy-makers, the CPI project included three multi-day Regional Team training workshops and fifteen community assemblies facilitated by the village-based Community Teams. The community assemblies were used as information sharing opportunities during which the teams explained the CPI project, listened to community concerns, learned what community members wanted from their engagement, and developed comradery and rapport amongst those attending. These assemblies played a significant role in keeping the CPI project locally visible, making and sustaining community interest, and preparing people to be willing to change behaviors they were accustomed to practicing. The assemblies built on one another, each increasing village interest in, and stimulating local conversation about, the project. The three multi-day workshops created coherent and functioning teams representing distinct areas of specialization, cultural background, and academic and professional training.

Essential to the application of the CPI model is the following sequence of project planning and development steps: 1) secure funding – the Ecuador cholera project was funded by USAID, but funding could have been sought from multi-lateral organizations, PVOs, governments, local and national organizations; 2) bring together funders and organizers to jointly identify the problem to be investigated and identify the key groups to be involved; 3) develop a clear and detailed work plan; and 4) select pilot communities in which to test, and then adapt, the model. This initial phase of the planning activity concludes in a three to five day start-up workshop during which the various stakeholders come to agree on the basic design of the overall project. And while the design may change in the process of implementation (as it certainly did during the Ecuador cholera project), the start-up workshop is when the roles and responsibilities of the participants are defined, a detailed timeline and implementation work plan are developed, agreements as to the goals and overall objectives of the activity are shared, establishment of rapport and trust among the participants is initiated, and existing data are presented (Whiteford 1999:9). In Chapter 3, we discussed how the CPI model was derived from what was called the CIMEP model (that continued to evolve after the CPI model was used in Ecuador). The CIMEP flowchart (see Figure 3.1, in Chapter 3) shows the activities that could be included in this first phase. Once these planning steps are agreed upon and completed, the field project begins.

Implementation of the CPI Project

The field project is carried out by three interlocking CPI teams. In the CPI cholera project, the Technical Team (TT) was made up of the national and international advisors from both the Ecuadorian Ministry of Health (MOH) and USAID. The Regional Team (RT) was composed of Ecuadorians from the regional or municipal government departments of education, health, environment, and transportation, and NGOs. The Community Team (CT) members were drawn from the affected communities themselves. The Technical Team oversaw the entire project. In addition to the Technical and Community teams, there were two state-based Regional Teams (one for the state of Chimborazo, another constituted from government offices in the state of Cotopaxi) who guided the project with the four Community Teams.

The Cholera Project – National Background

In Ecuador the large urban areas such as Quito and Guayaquil had been able to control the cholera outbreaks within eleven months of the first reported case. The outbreaks, however, were not controlled in the 20 townships in the states where we worked. These 20 townships were in five states – two along the coast and three in the highland mountains of the Andes. The three mountain states of Chimborazo, Cotopaxi, and Imbabura all had dispersed rural communities with inadequate access to sanitary infrastructure such as clean water and sanitation. In addition to being the three mountain states where the epidemic raged undiminished for the longest time, they were where the largest concentrations of indigenous people lived.

When the El Tor strain of cholera hit Ecuador in May of 1991, Ecuador quickly became the epicenter of what would become a widespread epidemic that moved through northern South America and into Central America, and would proceed to spread throughout the continent. By the time the epidemic began to subside in 1993, Ecuador had suffered more than 85,000 clinically diagnosed cases with almost 1,000 fatalities, and many more who were sickened and who died from the disease without a clinical diagnosis. Cholera continued to spread among the dispersed villages in the high Andes, with 80 percent of new cases occurring in the aforementioned 20 rural townships. The disease did not spread randomly; rather, it moved toward the most vulnerable populations and to families like Mariana's. Mariana's family lived in one of those 20 rural townships where the epidemic continued, and her

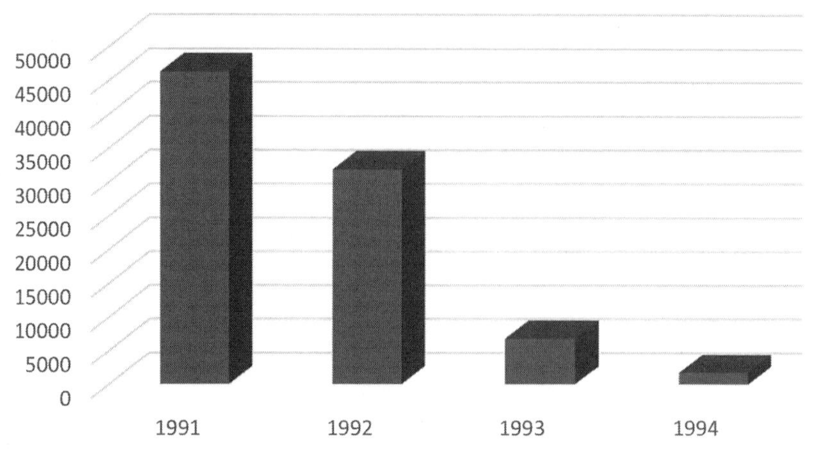

Figure 5.5. Cases of cholera in Ecuador, 1991–1994

family was similar to those of her neighbors in the highlands of Ecuador where, due to the lack of a reliable potable water supply and sanitation infrastructure, they were all potentially at risk of the disease.

It is neither surprising nor by accident that the epidemic lasted as long as it did in these states. While these states share a richness of culturally distinct traditions, festivals, beliefs, and practices, what marked them for the continued epidemic was, in part, their status as states predominately populated by indigenous Quechua-speaking groups. The states where the cholera epidemic continued shared poverty, exclusion from some avenues of economic power, and the continuation of institutional racism. Of these three isolated indigenous Andean states with the highest rates of cholera, we were invited to work with communities in the two states of Chimborazo and Cotopaxi. As Figure 5.7 shows, the rates of cholera were high in 1991 and 1992, and even though they were reduced by 1993, in 1994 these states still showed the continuing presence of the disease (Whiteford and Laspina 1996).

By 1993 the Ecuadorian Ministry of Health was confident that their aggressive programs of social (health) communication and hygiene education had effectively lowered and then controlled cholera in the urban centers, but they remained concerned about continuing outbreaks in the rural highland areas. The decision was made that to control the spread of the disease in these rural communities, a new model was needed. This new approach needed to be able to incorporate the cultural as well as the geographical constraints presented by the rural highland communities. The

Figure 5.6. CIA map of Ecuador showing the CPI states of Chimborazo and Cotopaxi in the center of the country

decision was made to collaborate with the Environmental Health Project, a USAID funded unit, and to employ the CPI model.

Once the CPI model was chosen, the personnel to staff the project were selected. The Technical Team was led by an Ecuadorian physician (Dr. Isabel

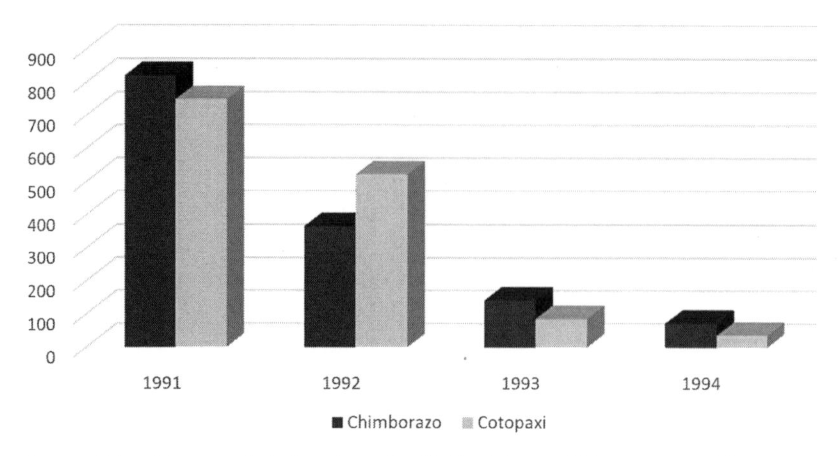

Figure 5.7. Incidences of cholera (per 100,000), states of Chimborazo and Cotopaxi, 1991–1995

Fernandez), who was also the Director of the Ministry of Health's Department of Community Health and Development, and another Ecuadorian, social psychologist and non-formal educator (Dr. Rosario Menendez) joined Dr. Fernandez, as did a U.S. medical anthropologist/public health scholar (Dr. Linda Whiteford). These three people comprised the Technical Team. Members of the TT then facilitated the selection of the Regional Team, which was led by the Director of the Chimborazo State Health Department (Dr. Sofia Velasco), who was also a physician. Dr. Velasco and Laura Amato (the nurse in charge of Community Outreach for the Chimborazo State Department of Health) were critically important to the project because they were known throughout the state, even in some of the small villages. Other members of the Regional Team were drawn from the Ministry of Education and were regional education leaders like Eduardo Gutierrez, while others came from governmental agencies in charge of water and sanitation. The Community Team membership emerged as village assemblies took place and the project was explained. We wanted community leaders to be involved, but not all community leaders were well-suited to the tasks of the Community Teams, and some people not traditionally considered for leadership roles by their age or gender were explicitly invited to participate.

Aims of the Cholera Project

The immediate goals of the project were to identify the beliefs and behaviors that were implicated in the continuing spread of cholera and to make recommendations for community-based involvement to change those

Figure 5.8. Regional Team member and his father

beliefs and behaviors in culturally appropriate ways. The long-term goals were to facilitate the development of local leadership that could help sustain those changes and to develop a replicable and sustainable global health model. Luckily, when people have an opportunity to learn how the soil on their hands could make their children sick, and how they can protect themselves and their families, they are interested.

> *"This project is very good because it showed us how important cleanliness is for health. The most important things I learned were to treat the water in the* bidón *[bottle] with chlorine and to clean and use the latrine to avoid illness and keep our families healthy. I want to teach my children these new behaviors and to help them help the community continue the new behaviors."*

In isolated rural communities, most without piped water and sewerage, personal hygiene is a luxury enjoyed by few. The most isolated of the project communities did not even have a health post or grade school.

Cholera Project States

In the first year of the epidemic, there were 46,320 cases of cholera reported in all of Ecuador at a rate of 43 cases per 100,000 people. Outside of the urban centers, the epidemic was most heavily concentrated in certain coastal and Sierra states. Chimborazo's (one of the Sierra states) case rate, for instance, was 86 per 100,000 people, and Cotopaxi's was 789 cases per 100,000 people. Compared with the national level of 43 per 100,000 people, it was clear that the two Sierra states of Cotopaxi and Chimborazo were in crisis, and within those states, the highest rates occurred in the indigenous communities (Whiteford et al. 1996).

Cotopaxi and Chimborazo were famous for the splendid and magnificent volcanoes after which the states were named, but also because the communities were located at high altitudes (about 9,000 feet above sea level), isolated from the urban centers of the country, and primarily rural. Wealthy Ecuadorian families in the capital often bought summer houses in the highlands because the beauty, cool weather, and quiet made the area a perfect escape from city life. However, for most of the permanent residents of the area, life was not so blissful. Most of the indigenous people in Ecuador live in these areas, many of them in the two states selected for the cholera project. It is not by accident that areas with a high percentage of their population being indigenous were also among the poorest and most marginalized regions in the country, thus rendering the people who lived there highly susceptible to disease.

In Ecuador, the majority of the indigenous groups live in isolated areas, experience high levels of poverty and infant mortality, and have limited access to employment, education, health care, public infrastructure (such as potable water and sanitation services), and other public programs

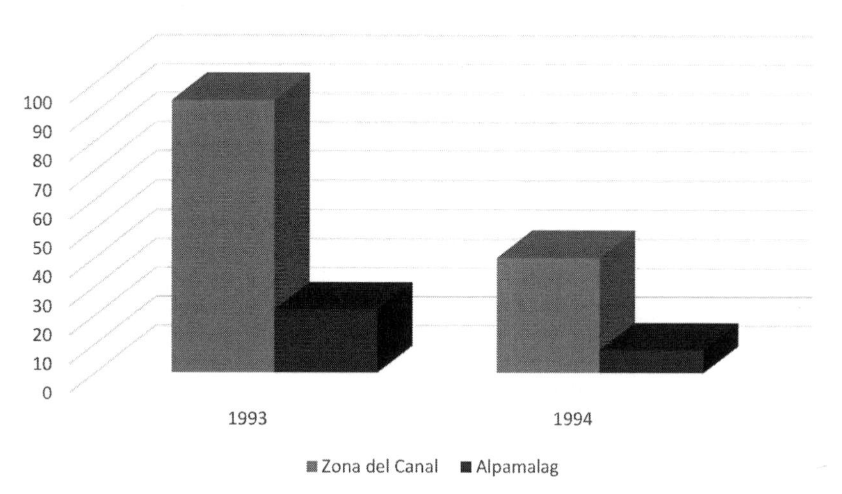

Figure 5.9. Cases of cholera in Cotopaxi project communities during the project

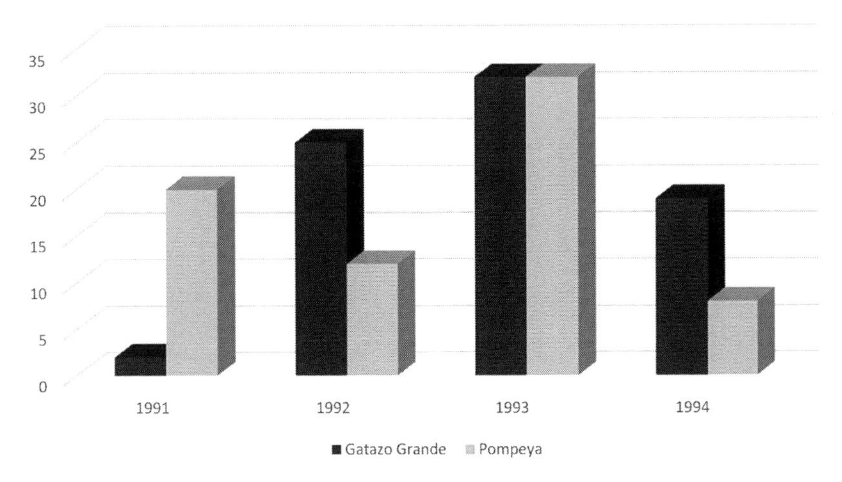

Figure 5.10. Cases of cholera in Chimborazo project communities during the project

(Whiteford and Laspina 1996; Whiteford et al. 1996). Unfortunately, it is a combination of these factors that creates the poor health status for the people in rural Ecuador, and this status, combined with a lack of sanitary living conditions, puts them at high-risk for contracting infectious diseases like cholera. Furthermore, a historical distrust between indigenous populations and outsiders creates cultural and structural barriers to timely and effective medical treatment for these diseases (Whiteford and Bennett 2005).

In addition to poverty and geographic isolation, these areas suffer from high levels of labor out-migration. Since the 1970s, Ecuador has experienced massive rural-to-urban migration from the rural highland Sierra states to urban centers such as Quito and Guayaquil. Men migrate out to work as laborers and are only able to return home a few times a year for ritual fiestas or short visits. Women, children, the elderly, and the disabled remain in the small Sierra communities. Women's work in the communities is particularly arduous for a combination of reasons: traditional gender roles place women in a lower status than men, and wives are expected to be subservient to their husbands. In addition, when men migrate away from home to find work, their home and community labor falls to women. Thus, women work the fields, tend the livestock, and work on communal labor crews, as well as care for the children, the home, and the elderly (Whiteford 2010). Though there are exceptions to the commonly encountered paternalistic family structure, the four communities (two in Chimborazo and

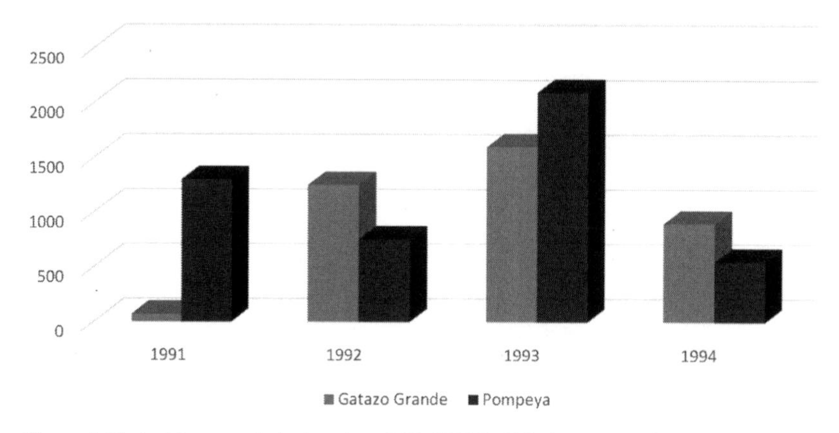

Figure 5.11. Incidences of cholera (per 100,000) in Chimborazo project communities

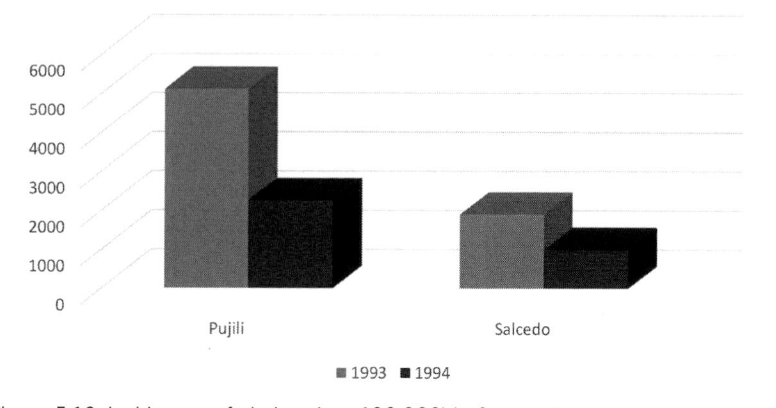

Figure 5.12. Incidences of cholera (per 100,000) in Cotapaxi project communities

two in Cotopaxi) that became sites for the cholera project reflected these traditional gender role patterns (Whiteford 2010).

The institutional inequality and structural violence legitimated in these Andean states was made evident in people's lives. Mariana, older and widowed, had more freedom than if she were young and unmarried, but she also had more responsibilities as she took care of her parents (until her father died), her own children, and the family fields, and once her aunt died, Mariana also helped to care for her aunt's young children. Alberto's life also reflected the traditional cultural expectations of his community. As a young unmarried man he was not considered for any roles in the groups that governed his community. He was considered to be too young, and not yet a householder, was not accepted in the community council, which was made up of older men, the majority of whom were married and with children. Because he was unmarried and still lived with his parents, Alberto was treated as though he was a child in the eyes of the older men. But at 20 years old, Alberto wanted to assume more responsibility for the welfare of his community, and while his age worked against him in the eyes of the older men, it made him more open-minded to change. This helped make him a leader in the cholera project and, later, in the broader community.

Poverty affects many Ecuadorians and shapes the lives of an even higher percentage of people in the rural areas. Sierra states, such as Chimborazo and Cotopaxi, have a greater concentration of poverty than other areas, receive a reduced amount of governmental assistance, and contain large numbers of indigenous groups with their distinct languages, cultural beliefs, and behaviors that further complicate distribution of the few resources allotted to these areas. In the early 1990s, in the states of Cotopaxi and Chimborazo almost 80 percent of all townships fell below the national average for levels of poverty and lacked basic sanitation services. Because of urban population densities and other political and economic factors, urban areas were where the wealthy and well-connected people lived, and the Ecuadorian government supported basic sanitation improvements in cities more often than in rural areas. Also, in urban centers it is easier and less expensive to provide services for more people than delivering the same services in the remote and isolated areas of the country (Whiteford and Laspina 1996; Whiteford et al. 1996).

Sanitation statistics clearly illustrate the discrepancies between urban and rural areas. For example, in 1994, in urban centers 75 percent of the people had access to potable water. However, this percentage dropped to less than 30 percent in rural areas (Whiteford and Laspina 1996). Waste disposal statistics were similar. In cities, 60 percent of the population had sewerage available, whereas only 10 percent of the people in rural areas had this same basic service. In Cotopaxi and Chimborazo specifically, upwards of 90 percent of people did not have access to potable water, latrines or sewerage services (Whiteford et al. 1996). Each of these deficiencies increases the vulnerability of indigenous populations and places them at a higher risk for contracting water-borne diseases like cholera.

Even in those communities where the water system infrastructure exists, often the water was not disinfected because insufficient funds were locally collected from the water taxes and so there was no money to buy the disinfectant. Therefore, even though the water may be piped, it may not be potable – a distinction difficult for many to remember when the water comes into the house and looks clean but is still not safe. When people do not know about 'those *bichos* (germs) you cannot see,' it is difficult for them to justify the extra cost of disinfectant.

Moreover, in these communities, as with many others in the world, people live in close contact with their animals (for instance, with *cuyes* [guinea pigs], chickens, pigs, or cows), often sharing a common shelter. Housing was often built from earth and clay and open to the surrounding environment. Roofs were often made of thatch, and floors of dirt. After the cholera project introduced soap as part of the hand washing campaign, we asked people about their use of the soap in both hand washing and dishwashing. In many of the homes people initially told us that they were using the soap, although we noticed that the soap rarely needed to be replaced. Given that the project was providing the soap, the cost of soap, we believed, was not the reason that the soap was not being used. So the teams began to ask more ethnographically in-depth questions about its use. Most people said that they were using the soap, just very judiciously. However, Mariana explained: "Pigs don't like the soap. If we use the soap, for dishes and to wash our hands, it stays in the water. And then the pigs won't clean up what we give them." Hence, we learned, the failure to use soap rested with the picky pigs, not with the people.

Inadequate food further stresses peoples' ability to withstand assaults by disease. In Chimborazo, 70 percent of the population lived in what the national government defined as poverty. The infant mortality rate was almost 50 per 1,000 live births, and 67 percent of the population was malnourished. While statistics for the state of Cotopaxi showed slightly lower levels of poverty than Chimborazo, they had similar high infant mortality rates (49 per 1,000 live births) and the same percent (67) of the population was malnourished (Whiteford et al. 1996).

Among the highland states with the highest on-going incidence of cholera, the team accepted the invitations to bring the cholera project into four communities where cholera was continuing to spread. The invitations came from the local leadership of the village council of each project community, as well as from local villagers. The Technical Team visited potential communities, explained the aims and goals (and methods) of the cholera project to local community members and local authorities like the village council, and asked if they wanted to invite the project into their community. If they did want the project in their community, they were also asked to become partners to the project and to work to facilitate the project however they could. Communities that were worried about who we were and what we might do did not ask us to bring the project into the community, and we did not. Other communities in which the leadership wanted the project, but in which local community members were too unsure about having outsiders in their midst, could not provide the assurance of partnership, and so we did not take the cholera project into those communities either. We situated the project only in communities that provided promises of partnership support and an invitation from the official authorities of the village, and in which we felt strong support from the local residents. In the end, we had many more villages that met those criteria than we had time, money, and personnel to support. Finally, we chose four villages with good support for the project, high and on-going levels of cholera rates, and variation among the four sites. And, with our four new partners, we began to translate the CPI model into practice to control the spread of cholera.

Most of the areas had been disproportionately affected by the cholera epidemic and had political leadership committed to understanding why and how the disease affected their communities. In the state of Chimborazo we accepted invitations to locate the project in Gatazo Grande and Pompeya; in the state of Cotopaxi we accepted invitations to bring the project to Alpamalag de la Co-Operativa and Comunidades de la Zona del Canal. All four communities were poor, indigenous, rural, isolated, and reporting new cases of cholera.

As Figure 5.13 shows, by the time we began to work on the cholera project, the number of cases country-wide had begun to decrease. However, when you look at the four communities in which the project was located (Figure 5.14), you see that the regional hospital was still treating significant numbers of cases of acute diarrheal disease from these villages.

Overview of the Four Project Communities

The communities in which the intervention occurred were similar to one another in many ways – for example, their mountainous ecology, the indigenous Quechua traditional cultures, and the distance from major urban regional hubs of commerce and trade. And yet they each had distinct individual personalities related to their particular history, geography, and economy. In the following pages, we introduce the communities, provide a brief overview of the spread of cholera in the regions, and invite the people in the communities to speak.

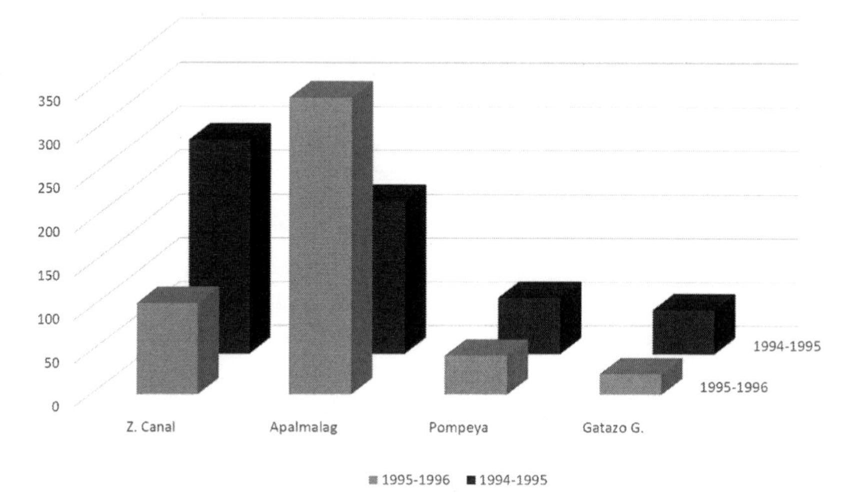

Figure 5.13. Acute diarrheal disease cases in project communities in the states of Cotopaxi and Chimborazo

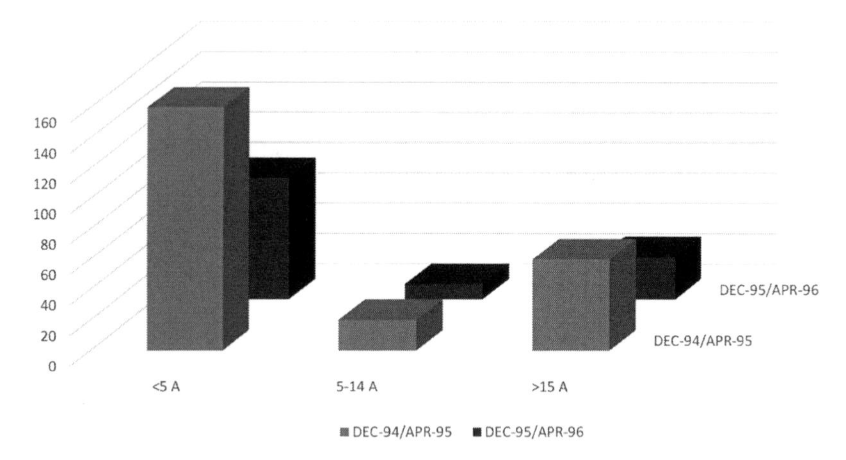

Figure 5.14. Acute diarrheal disease in Salcedo Hospital, by age and date

"You live with what you have. And, until you see how others live, your ways seem normal." Laura, the Regional Team member and Chimborazo community outreach nurse, was raised in a rural community in the highlands, and like the communities where the cholera project took place, her home was isolated and far away from any urban center. Laura could not have become a practical nurse without the help of family members who

Figure 5.15. Family transporting water

lived in one of the regional centers where the high school and nursing school were located. Laura had to leave her family and community and move into her aunt's house in town so that she could go to school. "If I had not moved into town, I would not know any life but the way my family had lived for many generations."

Laura was referring to communities like Gatazo Grande, Pompeya, Alpamalag de la Co-Operativa, and Comunidades Zonas del Canal, where traditional lifestyles dominated and where many people had little contact with the modern urban centers of Ecuador. 'Traditional lifestyle' means a variety of things depending on the community, but in general it refers to a lifestyle of reduced reliance on those accouterments we associate with modernity. In some communities it means little access to electricity, potable water, and sanitation. In the project communities it meant that most people did not own cars, but were dependent on public transportation where available, or relied on their horses or mules, but most often they walked. Most people lived near the fields they worked, although in some communities the fields were often far away and people walked distances each day to reach them. The communities had little access to health or educational institutions, having in some cases a grade school, or even a middle school. But no high school or health clinics. Those communities without piped water brought water in on the back of their animals, or carried it themselves.

By 1992, the overall cholera incidence was decreasing nationally; however, the rural areas still experienced much higher levels of the disease. As

Table 5.1. Cases of Cholera in Project Communities in Cotopaxi during the CPI Project

Year	Zona del Canal	Alpamalag
1993	94	22
1994	40	8
1995	0	0
1996*	0	0

(*through week 27. Note: Figures for 1991 and 1992 were unavailable. Source: DPS in Cotopaxi, Epidemiological Department)

Table 5.2. Cases of Cholera in Project Communities in Chimborazo during the CPI Project

Year	Gatazo Grande	Pompeya
1991	2	20
1992	25	12
1993	32	32
1994	19	8
1995	2	2
1996*	1	1

(*through week 27. Source: DPS in Chimborazo, Epidemiological Department)

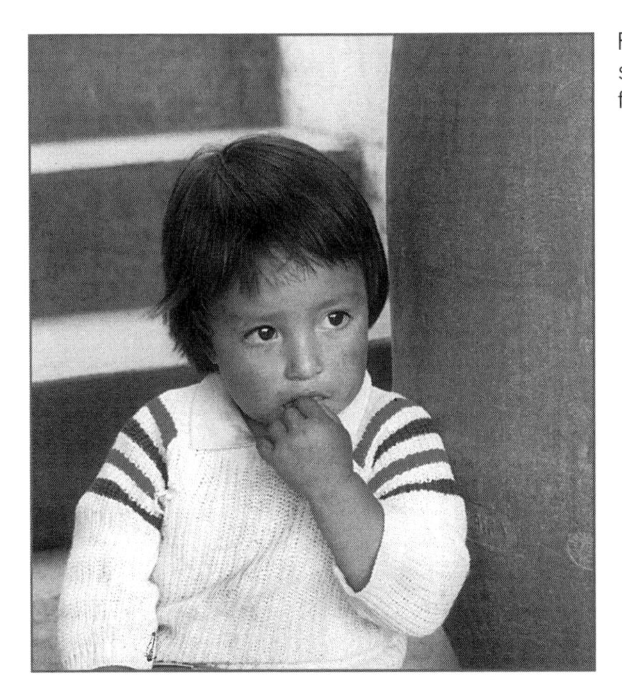

Figure 5.16. Child sitting next to a fifty-five-gallon water tank

Tables 5.1 and 5.2 show, the two communities we selected for the project in the states of Chimborazo and Cotopaxi, were communities where the epidemic continued through 1993 and 1994, which was much later than other regions that were by then largely cholera free.

Gatazo Grande

Gatazo Grande had a population of 2,000 people, distributed among 340 households. Quechua, a language spoken by indigenous groups in the Andes from Colombia through Ecuador to Peru and Bolivia, was the primary language spoken in Gatazo Grande, with Spanish spoken as a secondary language. Almost 80 percent of the people were nominally Catholic, and the other 20 percent belonged to fundamentalist Protestant churches in the area (Whiteford et al. 1996b). Corn, onions, potatoes, and other vegetables were locally grown on communal land, and animals such as rabbits, pigs, and guinea pigs were raised for private consumption and for sale in regional markets. Even though transporting items for sale to markets was difficult due to the associated transportation costs, poor roads, and an undeveloped distribution system, some community residents were able to do so, and provide additional income for their families.

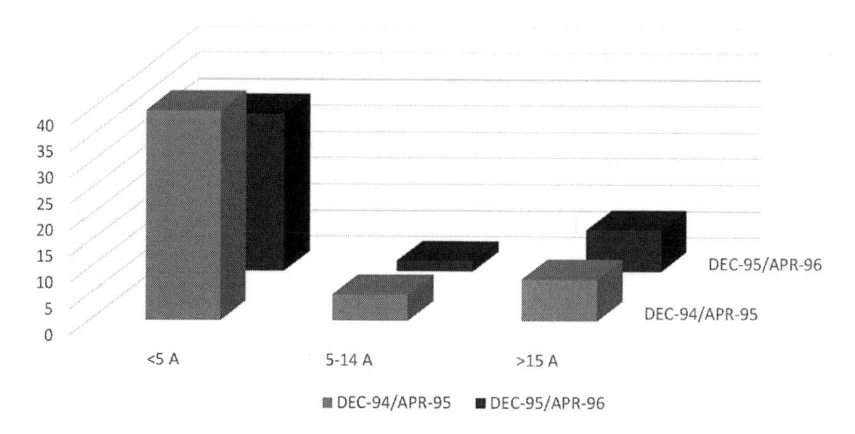

Figure 5.17. Acute diarrheal disease in Gatazo Hospital, by age and date (6–72)

Gatazo Grande's local government was centered on a president who was elected yearly. Two leadership organizations in Gatazo Grande were the Padres de Familia (senior, male members of the local families) and Comité del Agua (the Water Committee). The Padres de Familia focused on social development of children and worked with the local elementary school. The Comité del Agua was a more technical organization that oversaw the systems of community water distribution, maintenance, and user fee collection. Individual homes in Gatazo Grande had electricity, piped (but not necessarily potable) water, and latrines. However, the water was not consistently treated for bacterial contamination and the latrines were poorly maintained.

When we first interviewed people in Gatazo Grande in 1994, community members reported cholera, measles, and alcoholism as the most common diseases. In 1991 there had been only two cases of cholera, in 1992 the incidence jumped up to 25 cases of cholera, and in 1993 the increase continued with 32 cases reported. By 1994 the number of cases of cholera being reported had decreased somewhat but remained high with 19 cases.

Figure 5.17, while not based on clinically diagnosed cases of cholera, shows the high numbers of children under the age of five hospitalized with acute diarrheal disease. Acute diarrheal disease is often taken as a reflection of poor levels of clean water and hygiene. Gatazo Grande and Zona Del Canal are in the catchment areas for this hospital.

Pompeya

The second community in the state of Chimborazo involved in the cholera project was Pompeya. Pompeya was the most isolated and conservative of the four communities identified for the cholera project, and therefore provided a unique perspective of the challenges for a cholera intervention project in the most remote areas. Almost 1,500 inhabitants occupied 295 houses scattered throughout the community, leaving 45 houses that simply stood empty, their owners having died or moved away. Most community members were subsistence farmers, and communal land was used to grow potatoes, wheat, corn, and quinoa (a grain crop grown for its protein-rich seeds). Chickens, pigs, rabbits, sheep, and guinea pigs were also raised for feast day consumption and sale, often ranging uninhibited throughout the village.

Similar to Gatazo Grande, Pompeya also held yearly presidential elections and the position was one of honorary power used to organize communal work on community-based projects. Community labor groups, called *mingas,* existed within Pompeya. *Mingas* are a traditional form of communal labor used for cleaning and repairing shared resources, like a road or an aqueduct. When the community president called for a *minga,* for instance, to work on a road, all the men and women of the village, and children old enough to help, would be expected to participate in what is often a day-long back-breaking activity. And after a day digging trenches for water pipes, for instance, or planting potatoes on communal land, they each go home to do all the household, family, and livestock work which waits for them there.

More than once when we were returning from visiting teams in the rural areas, we stopped and picked up people on their way home as a *minga* was finishing their ten hours of labor. Everyone tired and dirty, but laughing and noisily happy to have a ride after a hard day of physical labor, was glad to squeeze into the little Jeep we drove. One time, after we had been driving straight into the cool midst and clouds of the mountains for an hour or so, we saw two small figures far ahead of us in the distance – they were two women in long woolen skirts wearing the felt hats traditional in the area and carrying axes and hoes. They were wearing rubber boots and had their hoes over their shoulders. As we got closer, we recognized Mariana and slowed down so she and her friend could squeeze into the Jeep. Mariana was happy to wave us down and climb in with us, making it six people in a four-seater Jeep along with hoes and axes. A tight fit, but a ride well appreciated.

The women had been out since before dawn, working with 80 other women from the surrounding villages on a communal labor project to repair water pipes damaged in the earthquake five months earlier. Mariana and the other women had gotten up early to feed their livestock before they walked for two hours to spend ten hours digging ditches. Now they were walking the two hours back to their homes, where they would check on their children and bring their livestock back down from the pastures. Then, they would prepare a meal and feed their families before they rested. It was a long and arduous day for all of them, but as Mariana said, "It is important work." Then she laughed and added, "And we women are *muy macho* [very tough]." And they were.

While all households in Pompeya were united by shared structural problems, such as the lack of piped water and latrines, they were divided by their religious beliefs. Forty percent of the population belonged to the Catholic Church and the other 60 percent self-identified as fundamentalist Protestants. Like so many other communities, Pompeya had experienced the divisiveness of religious non-governmental interventions that pitted one sector against the other. More than one religious group had been run out of town as they tried to provide their followers with some social and/or economic resource, while denying it to others in the community. Our teams were concerned about the chances for success of the cholera project in Pompeya due to its historic factions. We feared that the divisions within the community would sabotage our efforts to create a shared commitment to change. While we considered substituting another less closed, remote, and suspicious community for Pompeya, we concluded that Pompeya would be a good test of our model and methods.

Transmission of cholera in Pompeya was associated with social and economic relations, and we discovered these important transmission patters through in-depth interviewing with leaders in the community. For instance, in 1991 20 cases of cholera were reported; in 1992 the number fell to twelve; and in 1993 the number increased again to 32 reported cases. By 1994 only eight cases of cholera were reported. We proposed that cholera incidence in Pompeya was associated with the fact that many local men worked as labor migrants away from Pompeya in areas of high prevalence of cholera, such as Ecuador's major port city of Guayaquil. Every six months the men returned home to Pompeya to participate in community fiestas. Community feast days meant communal food sharing, preparation, and eating, often using hands rather than utensils. In a community like Pompeya with no piped or disinfected water supply, no latrines, and men returning from the port city of Guayaquil where cholera was endemic, we feared that these workers were re-introducing the disease to the community each time they returned from the coast. We recognized that the labor migration would not stop – these men were forced to migrate because there was little to no work available locally for poor, indigenous, uneducated, and relatively unskilled men, and thus we ran the chance that controlling the re-introduction of cholera into Pompeya could be beyond the project's abilities.

Alpamalag de la Co-Operativa

The third community was Alpamalag de la Co-Operativa, a small community in the state of Cotopaxi. Alpamalag also was an indigenous community with 428 people living in 120 households, all without electricity. Members of the community spoke primarily Quechua and used Spanish

Figure 5.18. Community members at CPI meeting

only as a second language. A small elementary school was located in Alpamalag; however, people had to travel more than two miles to the nearest health station, mostly by foot. For most of the year, the land surrounding the area was too dry to support any agriculture except agave, a cactus-like plant used to make tequila. Because of this, most adults, men and women alike, migrated out of Alpamalag in search of work and returned only for fiestas, leaving just the elderly and the young children at home much of the time.

Access to safe, potable water was difficult in Alpamalag. Piped water came into Alpamalag through an old water system that was frequently not working. Constant breaks in the pipes allowed soil to enter into the water system, carrying sticks, twigs, mud and other contaminants. Even when the pipes were repaired and the water system was working, the water was not potable because the water was not treated for bacterial contamination. Most people had water brought in on the backs of donkeys or in trucks. Some, who could afford it, would pay to have access to a spigot for water piped down from the mountains. More than half the community members had latrines, but these latrines were poorly maintained and the majority of them were not actually in use for excreta disposal, but rather for other purposes, such as storage.

Twenty-two cases of cholera were reported in Alpamalag in 1993, decreasing to eight cases in 1994.

Comunidades Zona del Canal

The fourth community involved in the Cholera Project was actually two small villages comprising 250 families. This area was called the Zona del Canal (Area of the Canal). As the name implies, irrigation canals surrounded the community and provided water to the nearby agricultural fields. The canals were not covered, and they passed by many homes. Because the canals were open to the air and because there was little organized trash removal, people tossed trash and human wastes into the canals. This was a problem because most families in Zona del Canal gathered their drinking water from these canals and used the water without treating it. In addition, a health post upstream was discovered to be dumping some of their waste (including from cholera patients) into the canals. Homes had electricity, and approximately half of the homes in the community had latrines as well. In 1993 residents of the Zona del Canal reported 93 cases of cholera, and in 1994 they reported 40 cases.

Figure 5.19. Waiting for the community assembly meeting to begin

The CPI Methods and Tools—Community Assemblies, Interviews and Observations, Perception Mapping, and Workshops

Community Assemblies

Throughout the twelve months of the project, community assemblies occurred often, called by the teams or the community. Initially, the assemblies were called to introduce and explain the project and invite community participation; later assemblies were called to invite volunteer community members for consideration to participate as a member of the Community Team. Once the project was underway, assemblies were called to share information acquired during the workshops, such as perception mapping, disease transmission, waste disposal, water handling, hand washing, and safe food preparation.

Interviews and Observations

Initially, members of the TT conducted interviews with community leaders, members of the community council, religious leaders, educators, and others referred to the TT as important in the community. Some, but not all, of those interviewed in the leadership interviews were invited to participate as members of the Community Team. Other residents, who had been excluded by traditional leaders because of their age, marital status, or gender, were explicitly invited to participate on the team. Structured interviews and observations continued throughout the project, and as the project proceeded, were conducted increasingly by the members of the CT themselves as they learned ethnographic techniques.

Across the four CPI communities, we focused on the three clusters of beliefs and behaviors identified as central to changing the spread of cholera: water treatment, hand washing, and food treatment. One of the most important ethnographic methods employed was visually monitoring water-handling practices, including water storage and re-use. The teams observed these practices within homes and at shared community spaces such as the common water spigot in order to study how people collected water, purified it, used it, stored it, and disposed of it. But what was 'common knowledge' (that water could carry bacteria and germs that were invisible to the naked eye) to public health workers was new and novel information to the majority of people in the project villages. Some villagers might have heard of germs, but certainly few adults knew anything about bacterial infections, and none appeared to be aware that bacteria could be carried in what looked like their 'clean' water. Doubly confusing was the fact that the water not only looked clean and clear, but villagers knew the water came straight from the snow-covered volcanoes they could see in the distance. So, they asked, how could the water be anything but safe?

Sometimes the local village teams laughed and giggled about the strange questions they had to ask while they carried out the survey for the project. The local team members asked their fellow villagers about beliefs concerning water, water handling, food preparation, and even about defecation. And sometimes these were difficult questions to ask and to understand why they were being asked. But as Mariana said, "This is serious business." For example, before the beginning of the project, many of the project participants rarely washed their hands, and almost never used soap when they did wash their hands or their dishes. They did not know why soap was useful, but they did know that it cost money they did not have. "We didn't know about the *bichos* [vectors, germs]. We could not see what made us sick; we didn't know to avoid them and that if we washed our hands with soap, it would help keep us from getting sick." If you don't know that germs exist, then there is no reason to use expensive things like soap for your hands or on your dishes. Or to use difficult to access resources, like water, needlessly. One villager told us how she used what water she had. In keeping with beliefs and behaviors in other water-poor communities, she re-used her water. The first use of the water was for cooking, the second was for rinsing dishes, the third use of the same water was for wetting hands, and finally the same water was used for the livestock.

The teams created and administered surveys about the three clusters of activities and engaged on observational practices in homes and in public places. Sometimes they were embarrassed to ask to go into their neighbors' homes. In the Andes, people lived public lives outside of their homes, but their homes were private and only for their family (and sometimes for their animals, as well). Therefore, entering private spaces for the cholera project was often a difficult endeavor and required a great deal of mutual trust and respect between those who lived in the household and others who participated in the CPI project.

Because the teams took the ideas that emerged from the various villages seriously, a sense of mutual respect was created among the three teams and between the teams and the communities. And participating in the CPI project changed both villagers and the team members. Each team member reported on ways in which he or she used the information learned; Eduardo and the other teachers, for example, took the lessons from the project into their classrooms and their students' homes. The quiet and solemn Susana steadfastly continued reaching out to young girls so that they also could participate in the project. The boisterous and no-nonsense Mariana continued to declare her right to health. And through involvement in the project, Alberto found his voice to assert his right to participate in the old men only town council, even though he was but a young man.

Perception Mapping

Informal community mapping (discussed in Chapter 3) became a popular and useful activity enjoyed by members of the community as well as by the CT. Community members were invited to share their maps with the teams (much to their amusement about what was included or left off each map), and the exercise became increasingly complex as the maps drawn moved from representations of geographic places to abstract concepts like bacteria and disease transmission. (See maps in Appendix C.)

Workshops

The workshops were designed to train Regional Team members in the skills and techniques of observation and interviews (ethnography), illness enumeration (epidemiology), disease transmission (germ theory), knowledge sharing (non-formal education), and leadership (capacity building). In addition, one aim of the workshops was to create an 'institutional memory' of the CPI model so that Ecuadorian governmental agencies could continue to teach and train others in the CPI methods (see Appendix A for greater details on the workshops). The Regional Team would then be responsible for passing on these skills to the members of the Community Team, who would in turn share the information with members of the local communities.

The first training workshop was held over a four-day period in which the members of the Regional Team were introduced to each other and the project aims, objectives, and methods. The workshop included team-building exercises designed to help the members of the RT learn to trust and communicate with each other and to create a sense of identity and cohesion within the team (Whiteford et al. 1996:7). The workshop was also used to train RT members in the skills and paradigms that the Technical Team wanted the Regional Team to be able to pass on to the Community Team and community assemblies. RT members were trained in techniques of cross-cultural and interpersonal communication and about models of communication, including paraphrasing, summarizing, question asking, and feedback. In this workshop, the RT was also guided on how to select the members of the four Community Teams, and how to train them to create perception maps of their community, to understand why such a map would be useful, and to be able to describe and analyze the data derived from the exercise.

The second four-day Regional Team training workshop built on the skills the teams had learned in the first training workshop. The second workshop was held after the RT members had held community assemblies during which they selected the members of the four Community Teams and trained them in the techniques the RTs learned in the first training workshop. The RT members reviewed with the TT their experiences with the Community Teams and the assemblies, and discussed their concerns, successes, and areas they wanted to improve or did not understand. In addition, in the second RT training workshop time was spent developing, adapting, and pilot-testing the research instruments. During this training workshop, RT members were taught to understand and practice methods of systemization and analysis of qualitative data, and, finally, how to create a work plan that included an instrument for monitoring community participation.

The third and final Regional Team training workshop was held for three consecutive days and allowed the RT members to share with

Figure 5.20. Regional Team member at CPI workshop

each other and with the Technical Team what had occurred when they introduced the concepts of qualitative research methods and developed a work plan with monitoring instruments with the CTs. The Regional Teams reviewed what they had learned in the previous workshops and how they had taught those skills to the CTs. Each community differed from the other communities in how they responded to these tasks, and each RT had to creatively respond to those differences. After the initial stages of learning communication and research methods, the teams were ready to embark on conceptualizing community-based projects, with budgets, work assignments, and a time frame. The last two activities in this final workshop were the development of ways to systematize the methodological processes employed in the CPI model and the creation of a work plan for community-based participative engagement and project monitoring (Whiteford et al. 1996:29).

When the RT workshops and the CT assemblies were completed, each village decided on what kind of engaged project to control cholera they wanted to develop. Each CT, with help from the Regional Teams, wrote a proposal for a community project, complete with aims, objectives, methods, time-line, and budget. These were then reviewed by the Technical Team and modified before being sent to appropriate regional, national, and international funding organizations. At the same time, the Technical Team selected a limited number of community-engaged projects to fund (but at least one per community). The provision of the funding was dependent on the community's ability to execute, sustain, and monitor the project and the resultant behavior changes. The funded community-engaged projects included the provision of safe water containers and an inexpensive way to disinfect the household drinking water supply, public garbage containers to be placed around the central square of each community (with requests for routine municipal waste removal), and local health fairs. All types of activities were funded, and in the next chapter, the results of the community monitoring are discussed.

Chapter Summary

- The CPI global health model is designed to be adapted to local and regional realities, and as such, culture, history, economics, and social cleavages need to be understood and taken into account.
- The Ecuador case study shows how institutional and historical racism are translated into the lack of provision of a safe and reliable water supply system or a sanitation system that heightens the risk of water-borne diseases like cholera.
- The four communities in which the CPI global health model took place demonstrate that even within a single country, variation exists.
- The methods of community assemblies, interviews and observations, perception mapping, and educational and team-building workshops are described.
- This chapter presents a case study of the application of the CPI model as it was employed to control the spread of cholera in four rural communities in the Ecuadorian Andes. The activities used in the intervention, along with the stages of application of the model, are presented in detail, along with brief descriptions of the communities and the people involved. While the actual workshop materials and tools are included in appendices attached to this book, the overall sequence of application is described in this chapter. In the next chapter, the outcomes of the intervention and its evaluation are discussed.

In-class Exercise: The Use of Paraphrasing in Community-based Cross-cultural Settings

Activity: To find alternative, but accurate, ways of communicating the same idea.

Goal: To communicate essential ideas without reliance on the use of jargon; for instance, how might one explain something like "biomedical information" without using those words? Think about what the underlying idea that is to be communicated, such as beliefs derived from Western medical science, and how that idea could be communicated using local concepts.

Methods: Allow the group to divide themselves into pairs, then:

1. Ask each person to write down a complex abstract idea as simply as possible. Then exchange what each has written with the other member of the pair.
2. Each person then tries to orally paraphrase the other person just said until they both agree it is an accurate representation of what was originally written down. See how many iterations it takes until both members of the pair agree to a final paraphrase.

Outcomes and Discussion

Outcomes

In the twelve months following the completion of the CPI project, community members changed the way they understood disease transmission and changed their behaviors. These behavior changes were encouraged through community meetings, and also through the provision of a five-gallon *bidón,* or water container. These containers were new, had spigots from which people could pour water easily, and had tight-fitting twist off caps that kept the water clean. Once community members understood that diseases such as cholera were transmitted in contaminated water, their attention to the family household water supply changed. Previously, water was kept in 55-gallon tanks, usually outside of the house, and often uncovered. More problematic was that people dipped their hands into these open containers, hands that may not have been washed since the person defecated. As villagers, especially those on the Community Team, were rewarded with these new, clean, and convenient *bidones* (and the chlorine tablets to disinfect the water), others asked to be involved in the

Community Participatory Involvement: A Sustainable Model for Global Public Health by Linda M. Whiteford and Cecilia Vindrola-Padros, 119–132.

project so that their households could also have a *bidón*. The Community Teams, with the assistance to the Regional Teams, then conducted workshops and trained other villagers. The new behaviors were encouraged and sustained primarily by the villagers themselves, although members of the Regional Team like Laura and Sofia continued to attend communities meetings. And Isabel, though she lived and worked in the capital (Quito) several hours away from the communities, monitored progress with the Regional Team Leader Sofia and both visited the communities.

After the initial twelve months of the CPI project, behavior change was evaluated both by the villagers themselves and by the Technical Team. The villagers on the local teams assessed whether behaviors in the community households had changed, and if the changes were being sustained. Both teams assessed behavior changes in the previously identified three clusters of behaviors: 1) household water treatment, 2) hand washing, 3) food preparation and dishes/storage, plus 4) the use of soap and disinfectants. The Community Teams came together to conduct personal house-to-house interviews and observations. Each team selected their own community member to lead the team and each team kept a weekly record of behaviors observed in his or her cluster of households. The final evaluation combined the village teams' records with observations and interviews from the Technical Team.

The overall results showed that even a year after the CPI project was officially completed, the four communities in which the CPI model was employed demonstrated:

1. An increase of 34% in households where water was treated with chlorine or was boiled.
2. An increase of 94% of households in which water was stored appropriately;
3. An increase of 42% of households in which dishes were washed with soap and clean water;
4. An increase of 27% of households in which people washed clean with soap and water;
5. An increase of 27% of households in which people washed their hands with soap and clean water after using the bathroom; and
6. An increase of 29% of households in which raw fruits and vegetables were washed in treated water.

The results showed marked improvements in all of the behaviors that had been targeted for change. Almost all households were storing their water in the *bidones* , and almost all households were storing their water safely. One-third more households than before the intervention were using chlorine or boiling their water; and there was almost a 50 percent increase in the use of clean water and soap for washing dishes. Equally important for the control of water-borne disease, there was a 30 percent increase in the number of households that washed their raw fruits and vegetables in treated water. We saw these continuing behavioral changes as marks of a sustained success, and so did the members of the communities.

Following the CPI project and its assessment, there was a significant reduction in new cases of cholera and no cholera fatalities in the four project communities. Table 6.1 shows the change from the baseline at the beginning of the project.

Table 6.1. Percentage of Water and Behavior Changes, Baseline and Follow-up Survey

Behavior	Baseline	Follow-up Survey
1. People engaged in food preparation wash their hands with soap and clean water.	25%	40%
2. After washing their hands, food preparers air-dry their hands or dry them on clean cloths.	20%	30%
3. After defecating or urinating, all people wash their hands with soap and clean water.	50%	77%
4. Hand washing is done in running water or in a container of clean water.	37%	46%
5. Dishes are washed with soap and clean, treated water.	15%	57%
6. Raw fruits and vegetables are washed in treated water before being served.	30%	59%
7. Excrement is disposed of in a toilet or cleaned latrine.	15%	72%
8. Children and adults defecate on open ground (fields).	69%	28%
9. Those who bury feces in open ground as a percentage of all residents.	16%	23%
10. All water used in household cooking, whether piped or stored, is treated chemically or by boiling.	36%	70%
11. Stored water is kept in small-necked, covered vessels and drawn through spigots or with a ladle used only for that purpose.	6%	100%

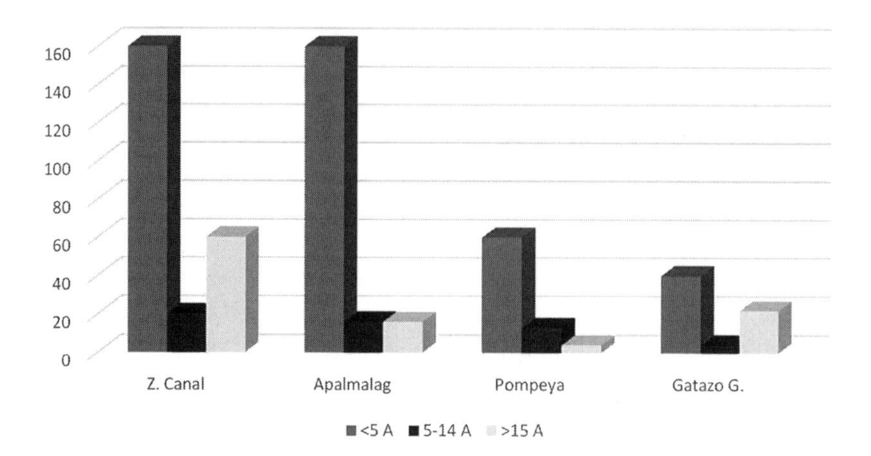

Figure 6.1. Acute diarrheal disease cases, by location and age, pre-intervention

The follow-up assessment also highlighted other positive findings regarding the incidence of cholera in the project communities. After the project was completed, the state of Cotopaxi reported 47 new cases of cholera, but none of these cases occurred in the project communities of Zona del Canal or Alpamalag de la Co-Operativa in that state. Similar results appeared in the state of Chimborazo, where 309 new cases of cholera were reported, but only four of these cases occurred in the two CPI project communities of Gatazo Grande and Pompeya.

By the end of our data gathering in 1996, only ten cases of cholera were reported in the entire state of Cotopaxi and none of these were in our project communities. At the same time, in the state of Chimborazo there were 59 new cases of cholera recorded with only two occurring in Gatazo Grande and Pompeya. Within the two years following the end of our project, none of the four communities experienced a single cholera fatality. Figures 6.1 and 6.2 show pre- and post-intervention cases of acute diarrheal disease in the project communities (Whiteford et al. 1996).

Discussion

The CPI project was successful in a number of ways, not the least of which was that there were no deaths attributable to cholera in the four project communities. The data suggest that villagers incorporated the messages into their daily lives. As a result, they broke the transmission of the cholera vibrio by boiling unsafe water, terminating the disposal of feces into water

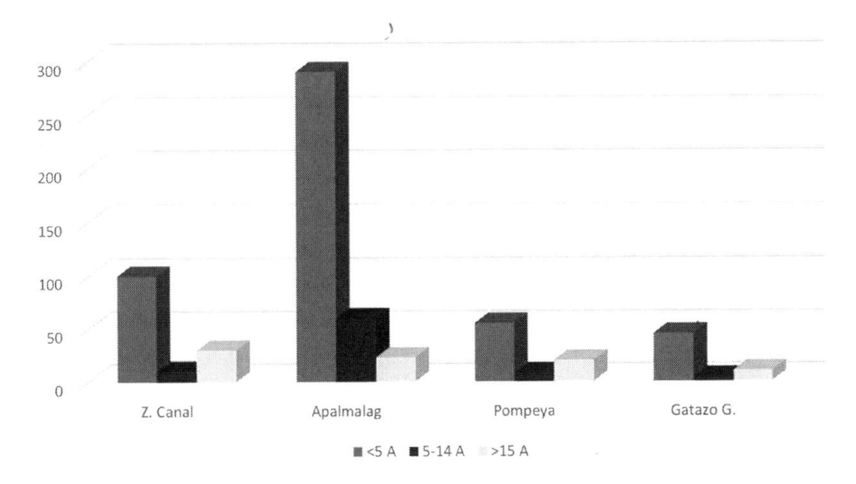

Figure 6.2. Acute diarrheal disease cases, by location and age, post-intervention

sources meant for drinking and bathing, storing their water in protected containers such as those supplied by the project, and washing their raw foods with potable water.

The project not only successfully mitigated the effects of a deadly epidemic in some of the remotest communities in Andean Ecuador, but also made an impact on the lives of those involved in the CPI program through education, training, and community involvement. The ethnographic, epidemiological, and educational training provided to the members of the Regional Teams facilitated the promotion of one of the members to the position of State Health Director, while others trained in community participation skills and techniques moved into better positions in the Department of Education. Eduardo, for instance, became a teacher in the Superior Politecnica School in Chimborazo. While members of the local teams became leaders in their own communities, members of the Regional Team often found professional advancement, within the government or, in some cases, employment with international NGOs, as a result of the skills and experience they acquired by participating in the CPI project.

Yet, the real supporters and beneficiaries of the cholera project were the communities themselves. As Mariana pointed out:

> *"Everyone's hands were dirty all the time; none of us knew it was important to clean them or to keep our water clean. I work with my hands all day – I plant and weed my garden, I feed the chickens and guinea pigs, I get*

> *food for my children. I work all day without washing my hands because I didn't know they could get us sick and I wasn't around any clean water or soap for washing. Now I know about germs and know that if I don't wash my hands, I might give the germs and sickness to my children. Now I use the water in the* bidón *and a little soap to wash my hands."*

And while Mariana still did not have piped water in her home, she had her water container and knew its worth and how to maintain it.

We knew from the beginning that these communities were cut off from many services provided to people in the other parts of the country. But we did not fully comprehend the extent of the isolation they experienced in their communities – Pompeya in particular – until we returned for the evaluation phase of the project. Isolation, in combination with poverty, creates a world in which tradition-based behaviors take on a powerful role and change is more than just a challenge to the ways things are done. Change becomes a challenge to the social order and to community expectations. The CPI project's provision of water containers allowed families to learn how to chlorinate their water and how to protect the water. Regional Teams leaders believed that people in the communities would take care of the water *bidones* if their potential worth to the family's health could be demonstrated. Pompeya, the most isolated, conservative, and divided of the four project communities, consistently showed lower response levels, less village participation, and greater difficulties in integrating the project messages into daily life, as well as continuing rates of cholera. And finally, the people of Pompeya showed fewer behavior changes than were found in the other project communities.

Taking into consideration Pompeya's history of rejection of outsiders, be they the government, the church, non-profit agencies, and even the local health officials, the teams were not surprised at the low response rate from the people in the village. The teams recognized the powerful force of isolation and tradition, and recognized that Pompeya was a village still harboring anger from past intra-village violence and closed off from outside contacts. Religious differences further divided the villages from one another, and even the Pompeya village council was initially conflicted as to whether or not to invite the project into the community. Once the CPI project was invited in to work in the village, the project still faced significant barriers to generating and maintaining effective engagement with the villagers. The local Pompeya team was relatively unresponsive to the various activities designed to generate involvement.

The teams looked for ways to approach each of the communities, with particular attention to Pompeya, that would not be threatening, perceived as outsiders telling the residents what to do, or disrespectful of the high cost of community-donated time. They recognized that the participation of these resource-poor, rural, suspicious, and alienated villagers, whether by attending the educational workshops, the town meetings, or just answering the questions that team members posed, took time away from their daily activities. And their daily activities started before dawn each morning and continued until after dark each night. Women got up before the others in the house to make a fire to heat some black coffee for the man and prepare a lunch for him to take to the fields. She might then wake an older child and tell it to watch the younger ones while she tended to the livestock, taking them up into the hills to graze, returning to feed any small animals like guinea pigs, rabbits, or chickens, weed the garden, work in the fields, cook food for the evening and next day, and then at dusk, head back up into the hills to bring the cow or pigs back to the house for the night. She and the children also probably hauled water from too often distant places. They had no time for anything else.

In recognition of how little time people in these communities had to spare, the CPI teams brought activities to the communities at times convenient to the villagers, especially to those in Pompeya. The teams brought health fairs that were fun for the children and held them when the children could bring their parents, and team leaders like Isabel and Sofia arranged for Ministry of Health and Ministry of Education people to come and provide village members with new baseball caps and t-shirts, crayons and coloring books, as well as with information about health and education. Pompeya was so remote – both geographically and culturally – that most of the members of the Regional and Technical Teams had never come to Pompeya before, or had ever even heard of the village.

With the help of these extra attentions and innovative approaches, some of the women from Pompeya not only learned about the importance of safe, potable water, but took an innovative and brave action – they asked CPI project member Laura if they could have an extra *bidón*. They told Laura how village mothers had no alternative but to leave very young children at home daily while the women worked in the fields high above the scattered homes. Most families took the older children with them to the fields, leaving the youngest children at home. Neither option was safe or healthy. Taking young children into the fields is dangerous; they are exposed for long hours to the cold and damp of the high altitude. While parents work the fields, little ones can fall into drainage ditches, eat unsafe things they run across, or wander off and hurt themselves. Similarly, leaving young children in charge of a baby at home for a ten-hour day can be dangerous; they fall on embers from the morning fire; don't get enough to eat, and they are contained in a dark and damp dwelling.

Based on what they learned from the project, these women wanted to use two rooms in a deserted building in the village to make a place where women could leave their babies and very young children during the day while they worked in the mandatory communal labor *mingas* or in the fields. One woman from the village would stay with the children during the hours when the rest of the mothers were away, and that job was rotated so that no woman would lose too much of her time and labor. In this little village-based *guardaria* (child care center), the mothers wanted the children to have clean water and decided to share care-taking if Laura could get them a *bidón* for clean water for the little day-care center to use. When we visited the *guardaria* during the evaluation phase of the project, it was cold, crowded, and damp and had about twenty children under the age of four in it. But they had someone to watch them and give them some food during the day, and places for their naps. And, true to her word, Laura and her supervisor, Sofia, did provide did provide them a *bidón* and disinfectant so the children had clean water. The development of the Pompeya community childcare center exemplified the very skills that we hoped to see emerge from the project: initiative and partnership combined with local leadership to treat a locally identified need.

In addition, the Pompeya community childcare center used the information about hygiene and sanitation to teach the children about clean water, hand washing, and even brushing their teeth. To make the program work, they sought funds and support from members of the Regional Team, who secured donations of toothbrushes from the Department of Health and coloring books and crayons from teachers like Eduardo and the Department of Education. The children were safe and learning about clean water and hygiene. "I learned from the cholera project why it is important to wash my hands, to make sure my drinking water is chlorinated, and to keep my house and latrine clean. These ideas changed the way I act. Now I spend more time cleaning my house and myself."

During the follow-up assessment, villagers in each of the project communities showed us where the public garbage containers (purchased as part of the project) were still being used, and they took us into their homes to show us their precious water containers. In almost every home the water container had its own cloth to cover the *bidón* – "to keep it clean" Mariana told us. We were impressed by the sacrifice made by the families in order to have a clean cloth covering the household *bidón* and the respect the little *bidón* was given.

The CPI project participants primarily learned about water-borne disease and water and hygiene practices, but they also learned a great deal more. "We didn't want to do that first survey; we didn't want to ask people those questions about health because [we thought] they weren't important. We thought that we knew what people would say, and we didn't want to enter peoples' homes." When the project began, local team members were hesitant and a little concerned about having to visit homes and ask 'intrusive' questions of their neighbors' beliefs and behaviors. In all four of the communities in the project, local tradition kept people out of each other's houses. People did not go visiting, and if they did, they stood outside of the house to talk. Inside was for families and their animals. Privacy was respected, no one talked about cleanliness or personal hygiene, and certainly neighbors never spoke about defecation. And now they had to ask questions about these delicate issues, and they had to ask if they could observe (some) behaviors inside their neighbor's house.

Despite these hesitations, the local teams did conduct both the initial survey and the follow-up surveys and interviews, and they entered peoples' homes to observe behavior as part of the assessment. And what they found out by going into people's homes surprised them. "People like us did not talk about health. We live, get sick, and die." What surprised Mariana and other members of the local teams was how little people knew about how to protect themselves from disease and how willing they were to learn. "Before the CPI project came, none of us knew very much about how our dirty water could make us sick, and we didn't know how to treat it so we could drink and not get sick. Now we know, and we try to help others not get sick."

"We never went inside of each other's homes before. Because of the project we had to let our neighbors come into our very homes to visit and observe our behaviors." Community team member Alberto told us that he was surprised that getting his neighbors to let him inside their homes was not as difficult as he thought, once it was clear that their opinions mattered and were valued. Alberto continued, "One of the benefits of this project is that it made it possible for us to visit each other's homes. It allowed the community to open up and participate." Despite the initial wariness, the final success of the project was caused, in part, by the Community Team members going into their neighbors' homes and people of the community becoming more willing to open their lives to each other and participate as a community in the project.

Even the relatively reserved Susana offered her opinion:

> *"The cholera project taught us new customs and behaviors, especially about how to keep our community clean. When we received the project's water container, we learned many ways to change the way we acted. We learned to clean our hands, to bury the garbage, and to teach our neighbors to put chlorine in their water. [We also learned about] the importance of keeping latrines clean."*

While not all of the communities had latrines, people living in those communities that did found them so filthy that they refused to use them. The CPI project helped people understand why the latrines needed to be kept clean and how to do it.

The communities did change, and their members worked to reinforce those changes and maintain them. The assessment results demonstrated changes that were observed in each of the four communities. The communities were cleaner, people were proud of their neat homes and yards; we witnessed community members burying excrement. Furthermore, water was being kept in the *bidones* provided by the project, and people were using water from these containers to wash their hands, food, and dishes.

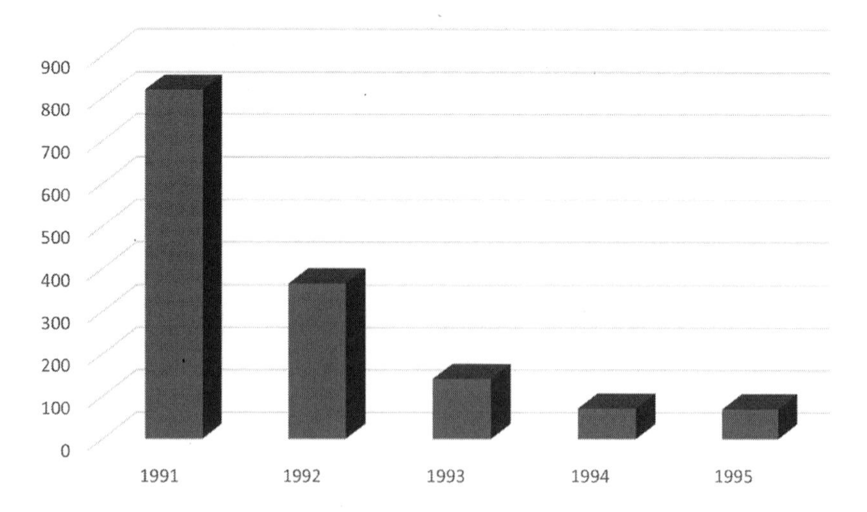

Figure 6.3. Cholera in Chimborazo, 1991–1995 (7–91)

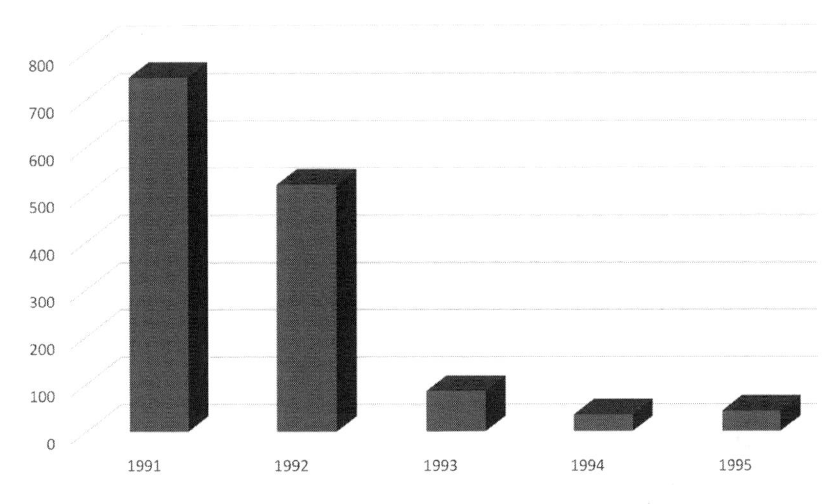

Figure 6.4. Cholera in Cotopaxi, 1991–1995 (7–91)

Latrines were being used properly for the first time in many places, and they were being kept clean as well. According to Alberto, "Parents may not worry about their own health, but they want their children to be healthy. Knowing about germs and how to keep our water clean, and ourselves clean, will protect the children." And Mariana added,

> *"The cholera project is so good that it deserves recognition for all the good it has done our community. It has helped us a lot. We now have a water container that helps us keep our water free from bad things and our children free of cholera. We learned how to change the way we think and act; the cholera project has been good for the community."*

As we were leaving, a Regional Team member asked one of the villagers from Zona del Canal to show us her water container. Initially, she declined. But as other villagers encouraged her to show us where she kept her water bottle, team members urged her to show us. She took us to her two-room home with dirt floors and minimal belongings. Her real pride, however, was across the unpaved road. It was her kitchen. It was a thatch-roofed building of two large rooms. She proudly pointed out the water container on a shelf, covered with a cloth to protect it. There was no furniture in the room, just a dirt floor and a cooking fire with herbs drying from the roof beams. While she was shy and reticent, when her fellow villagers asked her to explain to us why she was so proud, she said, "My house is small and my animals are important. I kept them in the kitchen with me to be safe, and then I learned that they can make me sick [to be] so close to the food." So she and members of the Community Team built her somewhere else safe to keep her animals. She proudly showed us the eight cages of guinea pigs and four of rabbits that had previously shared her kitchen, now safe in an enclosure in her yard.

The success of the CPI project extended far beyond the reduction of cholera in these four communities. It provided a venue for individuals and communities to engage their agency to make changes in a system where structural and everyday violence had limited their actions. A surprise, however, came from Mariana when she asked Laura, "But what will happen when our children marry and move away? Where will I get another *bidón* for them? They will want to take the *bidones* with them; what will we do?" Mariana's worry about her *bidón* (and about her children) provided insight into the importance of the little water container, the effect of understanding water-borne disease transmission, and the power of what villagers could do to protect themselves and their families. When Mariana asked about what she was going to do to keep her water clean and her household safe if she had to give up her *bidón* for her child who was getting married, her worries spoke eloquently about her understanding of the importance of the little water container and the ways to prevent disease. Laura assured her that until the government was able to supply the community with clean water, the Ministry of Health would continue to provide the little water bottles for them to use.

The outcome data and observations included in this chapter suggest that the CPI project was successful in the ways in which it was intended, and simultaneously, in other and unexpected ways. The changes in beliefs and behavior were what were sought; developing leadership capacity in the team members was certainly desired. The 'opening-up' of the communities, and the initiative and willingness to ask authorities for help in creating, for instance, the little day care center, were unexpected surprises. Lessons learned from the CPI project, and how they might be transferable to other global health problems, is the focus of the following chapter.

Chapter Summary

- Three of the four communities appeared observably more cohesive and open to community gatherings and activities than they had been when the project began.
- Gatazo Grande, Alpamalag, and Zona del Canal each appeared to have less loose trash and more of the residents engaged in both the CPI project and the new health behaviors.
- The exception was the village of Pompeya, but even Pompeya showed change as seen by the creation of the little community day care center.
- Each of the four communities benefitted from the cholera project, albeit in different ways. "The closeness we now feel as a community helps us provide for ourselves." Mariana went on to explain that people in her community (Gatazo Grande) had benefited from the community gatherings because they learned more about health, ideas about community participation, and their own community.

In-class Exercise: Evaluation of the CPI Project

Activity: Consideration of evaluations of community-based projects and the different types of data they employ.

Goal: Understand some of the different types of data and their uses in evaluations.

Methods: Analysis of the readings and discussion of alternative means of applications of evaluative methods.

1. Provide a description of the evaluation methods employed in the CPI project. What kinds of data did the local teams take into consideration?
2. The chapter states that the CPI project was successful in a number of ways. What were some of these ways and what kinds and sources of information were used to evaluate its 'success'?

Chapter Seven

Reflections and Lessons Learned

Reflections

In this book we have employed the CPI model to demonstrate how a medical ecology framework helps us understand the complex inter-relationships among biology, environment, and culture as expressed in the spread, and then in the control, of an infectious disease like cholera. So far we have focused on the mico-level of individual behaviors within the environment of the community. In this final chapter we take a more global and reflective approach as we show how the structural violence of history shaped which communities were at the highest risk of the disease, and the ways in which the communities responded. And we conclude with a discussion of the lessons learned from this approach in anticipation that the CPI model will be applied in other places around the globe.

> *"This project improved the level of education in the community and developed new leaders in the areas of health and sanitation. These leaders not only learned about health, but they put their knowledge into practice."*
>
> (Sofia Velasco, Regional Team leader)

Community Participatory Involvement: A Sustainable Model for Global Public Health by Linda M. Whiteford and Cecilia Vindrola-Padros, 133–147.

The story of the cholera epidemic reflects the history of political and economic marginalization of the indigenous population in Ecuador. Employing an analysis of the structural and everyday violence helped us identify how the indigenous communities were placed at much higher risk of water-borne disease than were non-indigenous or urban communities in the country. It showed us how social categories of exclusion were reified, reiterated, and repeated through time. Indigenous people in many countries are often systematically excluded from high-paying or high visibility jobs in government or in private industry; their communities are often in remote and difficult to reach places, and governmental infrastructure rarely provides access to health services, clean water, sanitation or education equal to that found in the non-indigenous communities. These institutionalized exclusions (structures of violence to the people they affect) are both so common and so long standing that often people are either blind to them or assume that they are 'the natural way things should be,' thus removing the initiative to change them. As Farmer (1997) has repeatedly shown, the institutionalized structures of inequality, such as the lack of health posts or permanent medical personnel to staff them, puts some groups of people at higher risk of suffering an early and unnecessary (or, as Farmer refers to them, "foolish") death than others. This is, indeed, a violence to those communities.

History is not passive, but continues to actively affect one's daily life and consume one's attention. Each of us is simultaneously affected by what has happened in the past – to our families, to our people, to our histories – and living in the moment (Fassin 2009). Some say that the scars of previous generations shape our sense of future possibilities and limit people's ability to even imagine a different kind of life. But as we saw in the CPI project, when the possibility of change was offered, for instance, for Susana to become a member of the CPI Community Team, against all odds she became an effective and engaged agent of social change. She challenged the rules (that women were not allowed places of authority in the village power structure) that even she thought were unbendable and immutable, "that's the way it has always been." But once challenged, change was possible. Likewise for Alberto, being allowed to participate in the CPI project became a life-changing experience, even though, as the village council elders said, "He is just a boy and should not be allowed to participate until he is married with a family." The CPI experience gave him the opportunity and encouragement to demonstrate his potential as a leader. The skills and experiences that Susana and Alberto (and all members of the Community and Regional Teams) acquired through the CPI project transformed their lives. Susana went on to attend a secondary school outside of the village and was heading toward nursing school, while Alberto graduated from a teachers training school and returned to the village to teach in the local primary school.

All cultures create inclusion/exclusion practices, and in Ecuador, the native peoples were often and routinely excluded from opportunities available to other members of the population. The areas of the country dominated by indigenous peoples have the highest levels of illiteracy, the overall poorest health status, the fewest employment opportunities, and overall, the earliest age of death. This situation is not the result of random factors, but rather the result of a history dating back to the Spanish colonial occupation of Ecuador and the subjugation of indigenous (and often rebellious) peoples. What resulted was a long, and often unexamined, history of institutional racism and structural violence. As we saw in Susana's observation about gender and leadership quoted earlier, one of the dangers of structural violence is that it renders the violence done to people invisible and also acceptable, and thus difficult to confront and challenge. The CPI model allowed people living in their own small villages to challenge some of the exclusionary practices in the name of controlling the spread of a deadly disease. The agency the participants acquired in that process radically changed lives, as some Community Team members, almost unexpectedly, took advantage of the new possibilities to challenge the status quo.

People on the Community Teams like Mariana were hungry to learn how to defeat the disease and to protect their children, while others like Alberto saw measurable advantages to working with outside groups, learning new ideas, and acquiring new skills. Susana began to find a way to help her community by quietly supporting the new ideas about health and hygiene among girls her age. Members of the Regional Teams like Eduardo saw the project initially as a way to advance their careers, but in time saw the experience as a way to enrich their lives and their ability to connect with others. The Technical Team saw the successful translation of ideas into action and lives saved, but also the development of an effective global health behavior change model. In short, all participants benefitted.

Reflecting on the enduring success of the CPI project in Ecuador, we ask: What lessons can be derived from the project methodology and from its analytic frame that can be applied elsewhere? What we observed in Ecuador was that people even in the most rural, isolated, and relatively closed indigenous communities were willing to change the ways they talked about health and illness and the ways they acted in order to curb disease transmission and to sustain those changes to foster healthier communities. In the four project sites people learned to recognize and prevent cholera and other types of diarrhea, and their communities experienced a new sense of solidarity, accomplishment, and an increased degree of openness to change. The Ecuador experience certainly is just one case, but the same model has been employed in Tunisia, Benin, and Bolivia with similar results. Being able to generate such similar and productive results suggests that the CPI model may, indeed, be an excellent model for global health behavior change in multiple regions and with various foci.

Lessons Learned

The seven lessons learned from this application of the CPI could certainly be expanded, but these were the most evident and significant:

1. The Ripple Effect
2. The Power of Follow-up
3. The Strength of the Second Generation
4. Widening the Net
5. Developing Local Leadership
6. The Synergistic Energy of Epidemiology and Ethnography
7. Building the Foundation for Transnational Public Health Initiatives

The Ripple Effect

Because each community determined the design of its own intervention through community meetings and discussions, the interventions they chose were appropriate to the community they served. Members of neighboring communities watched the project communities as they gathered together to conduct the baseline health survey and discuss its results, and as they were provided multiple training sessions with the Regional Team in which villagers were taught new community organizing and data gathering skills and techniques. The neighboring communities were intrigued when the project communities received *bidones* and learned methods for keeping their water clean and their families healthy. The ripple effect was evident when the surrounding communities began to ask how their communities could have the project come to them. As neighboring communities asked for the skills, education, training, and the water containers,

Regional and Community Team members streamlined the original process and replicated as much of the project as they could in those communities. Since the funding was no longer available from international sources, new communities had to generate funds either locally, or in some cases, the Ecuadorian government was willing to help pay for educational resources to enable the ripple effect to take root in the new communities. As the surrounding communities organized for health, they also challenged regional authorities to provide better services to their villages. And as they sought ways to fund their health projects and purchase the *bidones*, they reached out to governmental and non-governmental organizations for assistance. In short, village leaders expanded their scope of authority as they sought, and sometimes demanded, improved services from the local and regional government.

Neighboring communities adapted the methods and ideas from the original CPI model, and they extended the focus beyond improving access to clean water and sanitation and controlling cholera to other areas central to their own needs. When the project methodology was replicated, as for instance in new communities in the state of Chimborazo, instead of a focus on human health (e.g., health and hygiene), the community leaders used the CPI methods and applied them to community empowerment through an agriculture and animal husbandry project. The animal husbandry project was designed to promote a livestock vaccination program. The village leaders applied the CPI model ideas to educate livestock owners about the need to vaccinate their cattle. If the owners complied and joined the vaccination program, they would be eligible to receive a supply of iodized-salt for their animals. Salt blocks or salt licks were recognized by the farmers to improve cows' health and were locally considered desirable to acquire.

As we saw earlier, people in the high Andes live and work closely with their animals. Guinea pigs, rabbits, and cows often live within the shelter provided by the home. Families depend on their livestock not only for labor and income, but also for companionship. In the salt project, one farmer who brought his cow in to be vaccinated was very concerned about his cow. While he wanted her to have the iodized salt, he didn't want her to have to feel the pain of an injection. But he also knew that he could not get one without the other for his cow, Butterfly. "She cries," he said. And he cried also; but the cow got the vaccination and the farmer got the salt block for her. And the project succeeded in achieving both of its goals through outreach and education, and a little leverage of salt and tears.

Still another 'ripple' came as Community Team members from Alpamalag, took the lessons learned at the community level out to a new target audience: "We can teach our children to wash their hands, drink only clean water, and wash their fruit. But we all eat food from the street vendors. How can we teach them [the street food vendors] these lessons?" To teach the street vendors about sanitation and hygiene, women from the community set up a stall in town in which they prominently displayed one of the project's water *bidones*. The women used water from the *bidón* to wash and prepare the food they sold, as well as to clean the plates on which they served the items. She continued, "Everyone bought food from us because they didn't want to get sick. The vendors asked us what we were doing that made people buy from us and not from them. And we told them." The woman's story demonstrates the power of learning in action and how it can shape future applications. Not content with only keeping their own households safe, the women of Alpamalag moved the story from the private domain of the household to the public domain of the street, thereby enhancing its reach many times over.

The Power of Follow-up

From the beginning of the initial CPI project participants were told that, while the formal cholera project would end after twelve months, there would be continuing follow-up in the project villages. We recruited the most dedicated – and willing – participants to help monitor the sustainability of the project and to keep records of those households whose members continued with their changed health behaviors. In essence, they became community health workers, perhaps slightly less formally trained than others who were part of established primary health care programs, but nonetheless they worked to enhance and maintain the health of families in their villages. To our pleasure, we discovered that the Ecuadorian government was willing to support these community-based changes with new resources. The fact that Technical Team Leader Isabel was so well known and respected in the national Ministry of Health, and that she was willing to continue to dedicate her time, made others in the ministry look approvingly on the village activities and decide to financially support these new healthy behaviors. In addition, Regional Team members Sofia and Laura continued their active support throughout the second twelve months during which villagers' behaviors were monitored. The support of the national and regional governments by providing time for Isabel, Sofia, and Laura to continue to engage with the project communities was critical to the sustained outcomes. In addition, the creation of the cadre of village health workers that formed the core of the follow-up stimulated on-going interest and assistance from the Ecuadorian Ministry of Health.

In addition to the personnel support, material support included provision of small gifts in the form of simple, colorful, and easy to understand educational pamphlets that were distributed to the project communities. Both the health fair and educational booklets were part of the Ecuadorian Ministry of Health's overall strategic objectives, but the assistance to those specific villages was provided because the village leaders had become articulate spokespeople for the health of their communities through their participation in the CPI project. The new leadership provided by the community health workers brought their villages to the attention of the ministry in new and positive ways. And with Sofia and Laura continuing their direct involvement, even some regional non-profit health-oriented organizations that had not been involved in the initial CPI project made contributions and considered initiating some of their own projects in the villages. In short, the village leaders had discovered ways to make their needs known and to seek assistance from the larger outside community.

The Strength of the Second Generation

After the end of the CPI project, Regional Team members initiated participatory empowerment projects in their own work places. One group of Regional Team members who were teachers created their own health and behavior change projects. They replicated much of the original CPI methodology, but with modifications to fit their needs and resources. They were innovative in their design and responsive to their new audience and aims in their application. The teachers paired literate children in their classes with their illiterate parents to conduct household baseline health and attitude surveys, and then the teachers began to monitor the health-related behaviors in their schools and students' homes. In several schools, the school directors became excited about the project and encouraged the grade school teachers to create lesson plans that focused on health and hygiene, using hand washing and water purification as examples of desirable behaviors. With the assistance of Sofia, the State Department of Health helped reinforce the lessons by providing teachers and their students with pencils and other much needed school supplies to use in the school-based health and behavior change projects. In addition to the school-based teaching, outreach, and household change activities, one Regional Team member initiated a project during a mandatory month-long teacher-training program to educate all student-teachers in the epidemiology and ethnography skills used in the CPI project. When they graduated from their student-teacher status, the newly minted teachers were sent out to schools across Ecuador, taking ethnographic and epidemiological skills and techniques with them to their new classrooms and communities.

Widening the Net

As they observed the success of the CPI project and its spin-off projects, some NGOs and governmental agencies who had previously been unavailable or uninterested in participating in the initial CPI project became interested and engaged (often encouraged by the communities themselves) during the follow-up activities. This pairing of programs and sharing of resources strengthened each community, the participants, their local projects, and even the partner organizations. Even in Pompeya, the most 'unreachable' of the four project communities, the health fair hosted by the Regional Team had the benefit of creating new partnerships that, in turn, generated new resources for the community. The health fair brought representatives of a number of health-related agencies to the community, and they provided health-related talks and demonstrations. Previous to the initial CPI project, this community had been so resistant to outsiders coming into the community that even MOH people had difficulty accessing it. Following the initial CPI project, but particularly during and because of the follow-up, Pompeya allowed increased contact with the MOH (through the Regional Team members) for childhood vaccinations and other health promotions. Villages that previously had not had access to many of the health promotion projects sponsored by the state or national government, or by non-governmental organizations, became accessible to them through changes stimulated by the CPI skill and leadership workshops.

Developing Local Leadership

The CPI project provided skills, techniques, and opportunities to a variety of people, most of whom had never had such opportunities presented to them. In the previous chapter, we told the stories of Alberto, Mariana, Eduardo, Laura, and Susana and how they maximized their opportunities to become true community and regional leaders. The nascent leaders emerged from multiple cultural categories, including women whose social position often denied them access to power, young people and students whose age traditionally limited their access to community leadership, and others who did not fit into traditional community categories of leadership. While the CPI project focused on developing local and regional leaders who could facilitate household behavior changes around water and sanitation, the results were far more varied. In the follow-up communities, it resulted in a group of villagers who became non-formally trained community health workers.

Synergy: Structural Violence, Epidemiology, Ethnography, Non-formal Education, and Participatory Empowerment

The Ecuador case study is a story that builds on the strengths of public health and medical anthropology methods and clearly identifies both the participatory process and the role of non-formal education as keys to successful global behavior change health interventions. But the story is also about recognizing the political and economic connections that tied Ecuador to the global community, and that simultaneously isolated and excluded the indigenous people of the Andes from that larger global economic and political community. The degree to which the CPI communities were integrated into, or excluded from, the global economic structure became an explicit part of the CPI model through the incorporation of the structural and everyday violence framework. We embedded our analysis of disease in the history of race/ethnic relations in Ecuador, as well as in an understanding of contemporary global economic shifts. During the CPI project, the world underwent a shift toward a tightening of neoliberal lending restrictions which affected rural and dispersed communities like those in the CPI project by further decreasing their access to goods and supplies. In addition, during that period and as part of the same economic policy change, government subsidies to farmers, agriculturalists, and small-scale producers were reduced, leaving many communities with higher prices to pay. These policies reinforced the structural violence already existing that only increased the health and well-being disparities between indigenous and non-indigenous communities.

Historical and economic information were only two sources of data incorporated into the project. In addition to those data, the CPI model was grounded on information generated by ethnographic and epidemiologic methods and resulted in the detailed analysis of complex, culturally sensitive issues. One of the ways this combination of disciplinary methods and theories can be demonstrated is through community-drawn maps of high-risk areas. As described earlier and in more detail in Chapter 3, the mapping exercise served a variety of purposes: it brought the community together to engage in the actual outdoor drawing of the initial maps and generated community-based discussion since everyone had an opinion about the various maps that people drew. It also provided a non-formal, casual, and non-threatening venue to begin to discuss both the community and disease transmission. The exercise also drew on and combined information gathered by using an epidemiological analysis to identify community health problems requiring traditional epidemiological methods to describe the patterns of disease distribution, such as from clinical diagnoses on state and national health records. Then, the identified communities provided local definitions of concepts such as 'risk' and 'transmission' derived using traditional ethnographic techniques of open-end and focus group (the drawn maps) elicitation. The ethnographic focus was to elicit local perceptions of the determinants of disease. The strength of the approach is that information generated by ethnographic techniques gives a human face and cultural context to the population-based statistics generated by epidemiological methods. Each technique gives part of the cultural ecology picture, but neither one is sufficient on its own. It is the synergistic effect of the combination of these methods that makes the CPI model such a powerful tool for global health behavior change.

"The river makes us sick." "Walking through the cemetery can give you cholera." "The path that goes behind the school gets people sick." These were examples of the ecologically based illness stories we heard. People knew the physical ecology of their village, but no one identified the drinking water as something that might make them sick. No one mentioned that fecal materials on their hands might make them sick. No one knew that these behaviors were implicated in the transmission of cholera; most community members had never even heard of cholera, let alone were aware that it was cholera that was killing them. The CPI project used the participatory process to provide information about the disease and the routes of infection, and simultaneously to facilitate local identification of the medical ecology of local behaviors. The CPI teams opened discussions with the communities about why the community members thought that areas drawn on the map were places where people could get sick. And we learned that the ideas the community members expressed had a strong basis in a shared reality, even though it was difficult for those not from the community to recognize it. Learning and observing the ecology of the world that participants lived in is an important part of understanding their worldview, especially at the local level.

Building the Foundation for Transnational Health Initiatives

The roles played by local communities, regional or state governments, and national and international governments in this model provide an effective and tested foundation for 'scaling up' and could create the foundation for multiple transnational public health initiatives. The CPI model differs from other community-based participation models in that its focus lies in establishing relationships between the state and the civil society. In this model, changes in the community are not understood in isolation but are seen as requiring input and support from a wide range of stakeholders. Local needs are seen as the responsibility of community members as well as different levels of political and civil authority and other actors such as non-governmental organizations.

This shared responsibility for community-based problems is reflected in the interlocking structure of the three teams that make up the model: the Community Team, the Regional Team, and the Technical Team. The model uses a variety of engagement techniques – such as knowledge creation workshops, community applications, open community meetings, and change demonstrations – to bring these teams together so they can analyze the problems affecting the community, scale up issues outside of the community's control, and develop joint strategies to address them. By incorporating stakeholders from groups outside of the community, such as regional and national government authorities, this model seeks to guarantee the sustainability of community-based changes.

Following the successful control of the cholera epidemic in the project communities in Ecuador, the CPI model was applied to another community-based education and intervention project to help villages and towns in rural Bolivia reduce infant and childhood diarrhea. Because of the diagonal integration required in the CPI methodology, groups of people representing local authorities, regional and national institutions, and health-based NGOs all had to be committed to the project. In order to demonstrate both the effectiveness and functionality of the model, the funders brought 25 people representing the Bolivian stakeholders for a three day visit to the project communities in Ecuador. That trip served as a crash course on the costs, methods, and results of using the CPI model. By the time they returned to Bolivia, the stakeholders were ready to engage in the initial discussions of how to bring the model to their communities in southern Bolivia, along with the frozen shrimp that many representatives of landlocked Bolivia brought home from coastal Ecuador for their families to enjoy.

In addition, the model has a demonstrated history of success in countries as distinctive from each other as Benin in West Africa, Tunisia in North Africa, and Ecuador and Peru in Northern South America, and it has been successfully employed in a variety of topical applications, such as health behavior change, animal husbandry, teacher education, development of municipal services in peri-urban areas, and the control of infectious diseases. Critical to the model's success have been the effective capacity-building, leadership training, and sustained change that are integral to the model and could provide a basis for transference in other locations and by other governments.

Conclusion

During the ethnographic and participatory portions of the research, we identified historical, environmental and political conditions as factors contributing to the spread of water-borne disease – conditions such as people drawing their water from open canals and the disposal of hospital waste into canals from which downstream residents drew their drinking water. The lives of the people in the project communities reflected processes of both historical and contemporary exclusion as they occurred among people who had few resources and little power. Poverty and powerlessness were the result of larger political decisions that resulted in some communities having conditions that fostered the spread of water-borne diseases. In both the states of Chimborazo and Cotopaxi water is naturally occurring and abundant, often coming from the numerous volcanoes throughout the area. But the conditions under which the water is transferred from source to ingestion were unprotected and easily contaminated by waste – human, animal, or institutional (e.g., hospital). Most often the water is not piped into homes, but rather is collected and stored by families. The lens of structural violence illuminated how the lack of piped water and sewerage systems, the result of years of neglect, caused abundant supplies of water to become a contaminated commodity by the time it reached the homes.

To understand why these communities did not have access to potable, uncontaminated water when there was a steady supply of clean water coming from the mountains that surround them, we employed a binocular gaze that included individual behaviors, their communities, and larger extra-community forces that shaped access to clean water. Understanding how the rural geography of the Andes and the historical record of prejudice against indigenous people combine to create inequalities and disparities between them and the non-indigenous urban dwellers positions the players in their contexts. The CPI model helps do that by focusing both on the micro level of individuals within their families and on the more macro level of the community and the global forces that shape the cultural life of the community.

Successful and sustained individual behavior change is at the heart of most global health projects. But individual behavior change can only be sustained when the surrounding cultural and institutional contexts are changed to help support those behavior changes. While the CPI project was successful in helping people change their behaviors, the project also tried to change the cultural and institutional contexts surrounding them. And the key to that change were the members of the Regional Teams. Drawn from State institutions, such as the Ministry of Health, Education, and Transportation, they made changes within their own workplaces which translated into policy and practice changes. The project workshops trained personnel in the Ministries of Health, Education and Well-being, and Transportation (along with members of NGOs) using non-formal education techniques, such as paraphrasing, ethnographic and epidemiological methods such as mapping out high risk areas, and understanding the disease transmission routes; all techniques taught and practiced during the community workshops were means to institutionalize the CPI model. Regional Team members, particularly in the state of Chimborazo, were deeply committed to the behavior changes they observed and the ways they could use the CPI model themselves. As Sofia, the State Health Director for Chimborazo, said: "I see the possibility of using this methodology in a variety of health identification and promotion activities, such as in agriculture, forestation, animal vaccination, and the development of community-based clubs." Regional Team members were surprised, impressed, and proud of the changes they observed in the frequency of high risk behaviors in the local communities where they were working. They wanted to reproduce the same changes they observed in other projects they were working on. At the regional and even national levels, the project stimulated multiple cases where the methodology was replicated, each with appropriate modifications. As a Regional Team member in Cotopaxi told the research team: "In the replication of the cholera project, I developed a more profound understanding of the process and its potential applicability to a wide variety of settings." As the Regional Team members learned by doing, so they taught others the same way. In addition to the project replications, parts of the ethnographic, participatory, and epidemiological methodologies found their way into other community and state activities, from several United Nations-funded projects to local cooperative community-based projects. The incorporation of community perception maps became popular and useful, and the paraphrasing skills acquired in the CPI project community empowerment workshops became incorporated into other train-the-trainer projects.

Chapter Summary

- The CPI is an effective and sustainable global public health model.
- The model is well adapted for resource-poor countries.
- The CPI model is proven effective for a variety of global health problems.

The CPI was an effective model at the individual and community level, but how can institutions be transformed to help sustain those changes? The results from the research in Ecuador, and other parts of the resource-limited world, suggest that to be sustained, the local level changes in beliefs and behaviors must be supported and validated by other entities, be they NGOs, the national government, or third sector organizations (L. M. Whiteford and S. Whiteford 2005). Without that continuing support, the burden of change may become too heavy to be borne alone and those behaviors too costly to sustain.

In-class Exercise: Expanding the Use of the Global Health CPI Model

Activity: Discuss the lessons learned and long-term effects of the implementation of the Global Health CPI Model.

Goal: Have class participants articulate how the model could be applied in other cultural settings.

Method: Individual analysis and class discussion of the following ideas:

1. Identify characteristics of the CPI Model as described in this book that are unique to the Ecuadorian cholera epidemic.
2. Identify what are the global and generalizable skills and methods that can be derived from the Ecuadorian example.
3. Identify other diseases to which the CPI Global Health Model could be applied.
4. List the advantages of the CPI Model over non-community-based models.

Appendix A

Brief Overview of Workshop Objectives, Contents, and Products

First Training Workshop

February 21–24, 1995

Objectives

The workshop, lasting four consecutive days, had the following objectives:

1. Introduce the Regional Teams (RTs) to the aims, goals, and objectives of the project, and actively involve them in its development.
2. Train the RTs in how to work in collaborative teams that cross disciplinary boundaries, and create a sense of identity and cohesion among the teams.
3. Transfer knowledge and skills to the RTs so that they can replicate them in the formation of the Community-Based Teams (CTs).

Content

The content of the workshop was as follows:

1. Discussion of the objectives and process of the Community Participation Intervention model.
2. Techniques of cross-cultural and interpersonal communication, models of communication, including paraphrasing, summarizing, question asking, and feedback.

Community Participatory Involvement: A Sustainable Model for Global Public Health by Linda M. Whiteford and Cecilia Vindrola-Padros, 149–154.

3. Strategies for working in communities and the identification of community leaders.
4. Methodologies and techniques for creating perceptual maps.
5. General information about cholera.
6. Criteria for the creation of the CTs.
7. Procedure for the development of action plans.

After the workshop, RT members were charged to apply the skills and techniques they learned in the workshop. The first task was *formation of the CTs*. To do this, the RT assembled members in their communities to explain about cholera and the CPI model. After an explanation and a chance for members of the communities to discuss issues, volunteers were sought and the CT formed.

The first task for the CT was to create a *perceptual map of the community*. The perceptual map exercise was designed to fill two sets of needs: to provide RT with insight into how CT members perceive their communities and to increase CT members' awareness of their own communities. Then community assemblies were held at which the perceptual maps were displayed. The individual perceptual maps became a focus of animated discussion of cholera risk factors in each community. Each community then combined the various maps into a single community map to give to the RT, which in turn made an enlarged (and often colorful) rendition of the map to return to the community (see Appendix C).

The perceptual map became a physical manifestation of the community's perceived environment and sites of health risks. The CT was asked to place on the map all of the items and places which might contribute to disease transmission. It was emphasized to the CT members that the map was to reflect their ideas about the nature of their environment and its health risks. Some maps were drawn in great detail, showing garbage dumping areas, irrigation canals, animal containment areas, latrines, and even animal defecation. Other maps were drawn with less detail but included common sources of disease transmission, such as local food vendors.

While each was unique, the maps illustrated common perceptions about the community environment, high-risk areas, and sources of contamination. The process of creating the maps focused community interest on the project, encouraged discussion, highlighted the role of the CT, and provided a tangible result of community work. In addition, the maps became a point of pride to the communities and a reference point for CT members in later discussions of disease transmission.

Second Training Workshop

April 10–13, 1995

Objectives

The second workshop was a four-day training session, during which members of the RT accomplished the following objectives:

1. Reinforce the collaborative spirit among the members of the RT for the continued use of the CPI model.
2. As a result of perceptual maps, identify high-risk behaviors in transmission of adult diarrheas and cholera.
3. Learn and apply ethnographic methods and observation instruments introduced and discussed at the workshop.
4. Become familiar with techniques for open-ended interviews and their application.
5. Review and adapt the interview guide to local needs, incorporating local terms and cultural beliefs.
6. Field test the interview guide (after observations/practices at the workshop).
7. Understand and practice methods of systematization and analysis of qualitative data.
8. Create a work plan that included an instrument for monitoring community participation.

At the end of the second workshop, the RTs were able to draw insights from the perceptual map exercise. They discussed advantages of cross-cultural interpersonal skills, such as paraphrasing ideas and using feedback to clarify information.

RTs also field-tested the open-ended interview and observation guide. Most members of the RTs had previous experience with closed-ended survey research instruments, but none had experience with open-ended questions and qualitative data.

This proved to be an important element in the project, since the RTs were willing to experiment and change their ideas about research procedures and to see the validity of community responses. The combination of ethnographic and epidemiology field techniques challenged and rewarded the participants. Before they left the second workshop, RT members had to create a work plan for the tasks to be completed before the next workshop. They also developed a guide for organizing and analyzing both quantitative and qualitative data.

The single most significant task undertaken in the second workshop was field-testing the interview guide and questionnaire. RT members had to be comfortable with and knowledgeable about the research instruments, since they would have to show CT members how to use them. The Technical Team (TT) prepared the research tools, but it was up to the RTs to modify and adapt the instruments to their own understanding. The final instruments used in the communities were further changed to reflect local understandings, linguistic terms, and concerns while maintaining the same general research foci. The instrument developed by the TT (see Appendix C) was generated by the project goals, previous research on cholera, and information necessary to understand local behaviors and beliefs. The final instrument used in the communities was the RTs' and CTs' own creation; the instrument developed by the TT was only a guide. Both RT and CT members reinforced the idea that the instrument could be used well only if it was adapted to the local contexts.

Several days of the workshop were dedicated to introducing, discussing, adapting, and field-testing the research instruments. While the instruments were not field-tested in the actual communities in which they would be used, the field-testing exercise proved to be invaluable. TT members accompanied small groups of RT members into a nearby community to test the instruments. The workshop group experienced greater rejection than the actual CT members eventually did, because each CT worked in its own community and neighboring homes, while the workshop group was testing the instruments where there were no established ties. Once the workshop teams returned from the community, the experience was discussed and the instruments modified. Each RT was responsible for modifying the instruments following a pre-test conducted by the CT in their own community. Thus, specific modifications were made for each locality.

Third Training workshop

June 13–15, 1995

Objectives

The third workshop occurred during three consecutive days and had the following objectives:

1. Analyze the community-based data on high-risk adult behaviors associated with cholera.
2. Discuss community knowledge, beliefs, and behaviors surrounding adult diarrheas.
3. Provide feedback about the ethnographic experience and the observation and interview guide.
4. Review qualitative data analysis methodologies
5. Conceptualize community-based projects as solutions to concrete problems identified by the community.
6. Learn and practice community intervention participation processes
7. Identify strategies for funding the community intervention projects.
8. Analyze and field-test participatory follow-up and monitoring of projects.
9. Systematize the methodological process of Community Participation Intervention (for use with other community-based problems).
10. Elaborate a work plan for the community-based participative intervention and project monitoring.

While each of the workshops covered a complicated set of topics, the RTs' understanding of the topics covered in the *third* workshop laid the basis for the possibility of long-term success of the proposed interventions. The workshop opened with a discussion of activities the RTs had completed with CTs and the results (and problems) which surfaced between the second and third workshops. Results were discussed and shared among the teams and the TT, and problems identified and, hopefully, resolved. During this workshop, RT members were trained in both data analysis and proposal writing. Data analysis was critical, because the proposal for a follow-on intervention must be based on an understanding of local behaviors, beliefs, perceived needs, and community willingness to contribute to the effort.

The third workshop was designed to last four days, but had to be compressed into three. Community data had already been sent to the TT for preliminary analysis. Further analysis was conducted during the workshop with the TT and the RTs. Following a discussion of the meaning of the results, the RT from each state worked on a hypothetical proposal for an intervention to gain experience in formulating aims, objectives, materials, costs, and time required for an intervention. This exercise provided the RTs with an understanding of how to conceptualize an intervention proposal and ways to deal with conflict resolution and problem-solving.

By the end of the third workshop, participants were required to develop a plan of action for the community-based intervention that included details of activities, a budget, and timetables.

Along with the plan of action, a guide for facilitating community involvement in determining an appropriate intervention was completed. The last product of the third workshop was the creation of a methodology for community-based monitoring of the proposed intervention.

RT members then returned to their CTs and passed along the information they had learned at the third workshop. They worked with the CTs to organize community assemblies in which to present the results of the community research and conduct discussions. Based on the community assemblies, CTs prepared a list of interventions that they and the community considered appropriate. The RT worked with each CT to evaluate the possible interventions and decide which was most feasible to consider, and then wrote a proposal for it. The ideas contained in each intervention proposal reflect the CTs' access to local/community ideas and the RTs' professional training and experience.

The proposals for community-based interventions were then sent to the TT, which read and evaluated each one. The TT then decided which interventions to fund and which ones to pass along to other funding agencies, both public and private.

Interview and Observation Survey Instrument

Observations in the Household

1. Observations Focusing on the Kitchen

1.1. Observation: What foods are prepared?

soup
rice
fruit drinks

1.2. Observation: How are foods prepared?

fried
parboiled/blanched
raw
boiled
reheated

1.3 Observation: Where does the water for cooking come from?

rainwater
treated, piped water
well or spring
untreated, piped water
canal or river

Community Participatory Involvement: A Sustainable Model for Global Public Health by Linda M. Whiteford and Cecilia Vindrola-Padros, 155–165.

1.4. Question: What kind of containers are used to store water?
- container with a small opening and a top
- container with a small opening without a top
- a jar with a lid
- large tank with a lid
- large tank uncovered

1.5. Observation: How does a person draw water from its container?
- from the tap in the container
- with a ladle used only for this purpose
- with another utensil (cup) used only for this purpose
- with whatever utensil
- with the hand

1.6. Observation: Are foods that are eaten raw washed beforehand?
- Yes
- No

1.7. Observation: Does the person cooking wash their hands with soap and water?
- Yes
- No

1.8. Observation and question: How do you wash your hands?
- with running water
- from a container of standing water

1.9. Observation: How does the person cooking dry their hands?
- air
- they don't
- on a towel
- on a used cloth
- on their clothes

1.10. Question: From where does the person cooking get the vegetables they use?
- their own garden
- the market
- the store (grocer)
- street vendor

1.11. Observation: Where is prepared food stored?

- in the refrigerator
- cupboard
- in a jar with a lid
- in a jar covered with a cloth
- in an uncovered jar

1.12. Question: Do you always reheat food before eating it?

- Yes
- No

1.13. Observation: With what do you clean dishes?

- water and soap
- water and ashes
- water only
- a cloth without water
- don't wash

1.14. Observation or question: How do you dispose of the dirty water?

- feed it to the livestock, like pigs

2. Observations Concerning Human Body Wastes

2.1. Observation and question: Do individuals wash their hands after defecating and/or urinating?

- Yes
- No

2.2. Observation: How does a person wash their hands after defecating and/or urinating?

- running water and soap
- running water, without soap
- in a basin of water with soap
- in a basin of water without soap
- in a container of water that is used for several things

2.3. Observation: How does an individual wash their hands?

- with running water
- in a container of standing water

2.4. Observation: How does a person dry their hands?

air
don't dry
with a towel
with a used cloth
on their clothes

2.5. Observation: How does one dispose of defecation?

toilet
in a latrine cleaned with water
in a dry hole latrine
by burying it
on the open ground, without burying

2.6. Observation or question: How often do people bathe?

once a week
every other week
less than twice a month

2.7. Observation or question: Where do people bathe?

a shower in the house
shower outside of the house
on the patio with water from a container
in a river

3. Observations Concerning Eating within the Household

3.1. Observation: What foods are being eaten?

coffee
fruit drinks
rice
soup
corn on the cob

3.2. Observation: What kinds of drinks are being consumed?

cola
oat based drinks("gruel") and fruit drinks
boiled water or juices made with boiled water
water or juices prepared without being boiled
maize liquor

4. Observations Concerning Washing Clothes

4.1. Observation: What is the source of water used to wash clothes?

rainwater
treated, piped water l or spring untreated piped water, untreated
river or canal

4.2. Observation: How is the dirty water disposed of?

on the ground
in a ditch
in an aqueduct
in the river
in the irrigation canal

5. Observations Concerning Outside Areas of a Household

5.1. Observation: What is the source of water used to water gardens?

rainwater
piped water
well
spring
canal or river

5.2. Observation or question: Do children or adults defecate in the fields?

No
Yes

5.3. Observation: Are animals enclosed in a corral?

There are no animals
Yes
No

5.4. Observation: Is there loose rubbish in the patio or garden area?

No
Yes

5.5. Observation: How is the majority of the trash disposed of?

buried
burned
in a ditch
in the river
in a canal
on the open ground
other

Observations of Non-household Situations

6. Observations of Food Vendors in the Street or Market

6.1. Observation: What foods are prepared?

fish
tortillas
rice
soup
sausage
potatoes

6.2. Observation: How are foods prepared?

fried
boiled
parboiled(blanched) or heated
reheated
raw

6.3. Observation: Are raw fruits and vegetables washed before they are sold?

Yes
No

6.4. Question: If the fruits and vegetables are washed, with what kind of water?

rainwater
piped water
from a well or spring
from a water vendor
from a canal or river

6.5. Observation and question: How do they (street vendors) wash their hands?

running water
in a container of standing water

6.6. Observation: How do they maintain the prepared foods?

very hot and covered
warm and covered
warm and uncovered
uncovered
uncovered, exposed to dirt, or in close proximity to the floor

6.7. Observation: How is the food sold?

very hot
very warm
lukewarm
room temperature
cold (with ice)

6.8. Observation: What are prepared foods served in?

disposable plate
on a wooden stick or skewer
on porcelain or metal plate, washed
on used office paper
in the hand
other

6.9. Observation: With what kind of water are dishes washed?

water and soap
water and ashes
only water
a cloth without water
unwashed

6.10. Observation and question: How are the dishes washed?

with running water
in a container of standing water

6.11. Observation: What kinds of drinks are served?

cola
oat based "gruel" or fruit drinks
water (boiled) or juices prepared with boiled water
water or juices not prepared with boiling water
maize liquor

6.12. Question: Are juices prepared with boiled water?

Yes
No

6.13. Question: How often are they prepared?

several times a day
once a day
once every few days

6.14. Observation: Have you observed street vendors washing their hands with soap and water?

Yes
No

6.15. Observation and question: How do they (street vendors) wash their hands?

with running water
in a container of standing water

6.16. Observation: Does the vendor get rid of the trash? (remove it from the vicinity of the food)

Yes
No

6.17. Observation: How does the vendor dry their hands?

air
with used paper or a towel
on a used cloth
on their clothing

7. Observations at Religious and Other Community Social Events

7.1. Observation: What foods are being eaten at the event?

rice
chicken soup
potatoes
fish

7.2. Observation: The foods that are served are:

fried
boiled
parboiled (blanched)
reheated
raw

7.3. Observation: How are the prepared foods maintained?

hot and covered
warm and covered
warm and uncovered
uncovered
uncovered, exposed to dirt, or in close proximity to the floor

7.4. Observation: With what are prepared foods served?
a ladle
a spoon
a cup
another utensil
the hand

7.5. Observation and question: What drinks are served?
cola
oat based "gruel" and fruit drinks
boiled water or juices prepared with such water
water or juices prepared with unboiled water
maize liquor

7.6. Question: Are juices prepared with boiled water?
Yes
No

7.7. Question: How often are they prepared?
several times a day
every few days
once a week

7.8. Observation: Is there a toilet or latrine near the event?
Yes
No

7.9. Observation: If there is a toilet or latrine, is it being used?
Yes
No

7.10. Observation: Is there a place to wash one's hands nearby?
Yes
No

7.11. Observation: With what type of water do people wash their hands?
running water
piped water
from well or spring
from a water vendor
from a canal or river

7.12. Observation: How do people wash their hands?
- with running water
- in a container of standing water

7.13. Observation: How is the food served?
- very hot
- very warm
- lukewarm
- room temperature
- cold (with ice)

7.14. Observation: In what are prepared foods served?
- disposable plates
- on wooden sticks or skewers
- on porcelain or metal plates, washed
- used office paper
- in the hand

7.15. Observation: With what are dishes washed?
- water and soap
- water and ashes
- water only
- a cloth, without water
- not washed

7.16. Observation: Have you observed the person serving the food washing their hands with soap and water?
- Yes
- No

7.17. Observation: How do they (food servers) wash their hands?
- in running water
- in a container of standing water

7.18. Observation: With what did the serving person dry their hands?
- air
- didn't dry on a towel
- on a used cloth or their clothing

7.19. Observation and question: After the party, what happens to the garbage?
bury it
burn it
throw in a ditch
throw in a river
throw in a canal
other – "wind"

7.20. Observation and question: Do people that live outside the community attend community parties and other community social events?
No
Yes

Appendix C

Perceptual Maps of Communities

Presented at Workshop III

June 13–15, 1995

Comunida de Pompeya

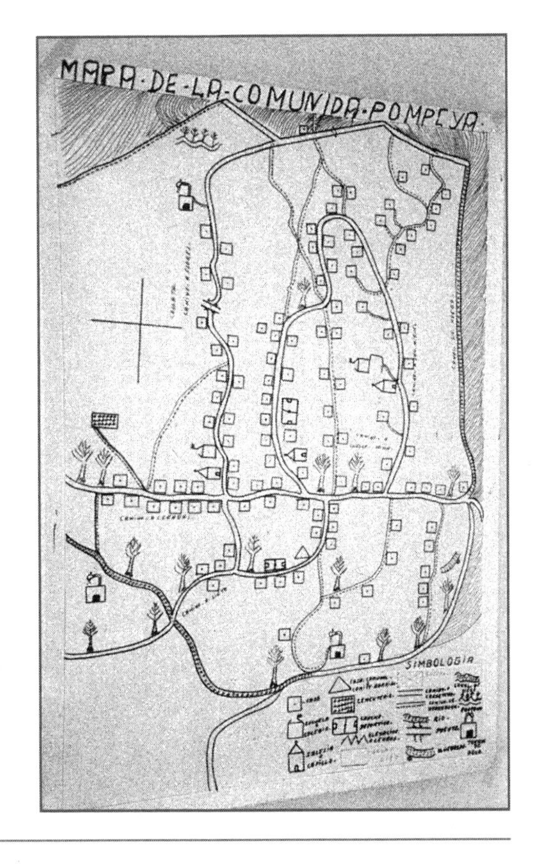

Gatazo-Grande

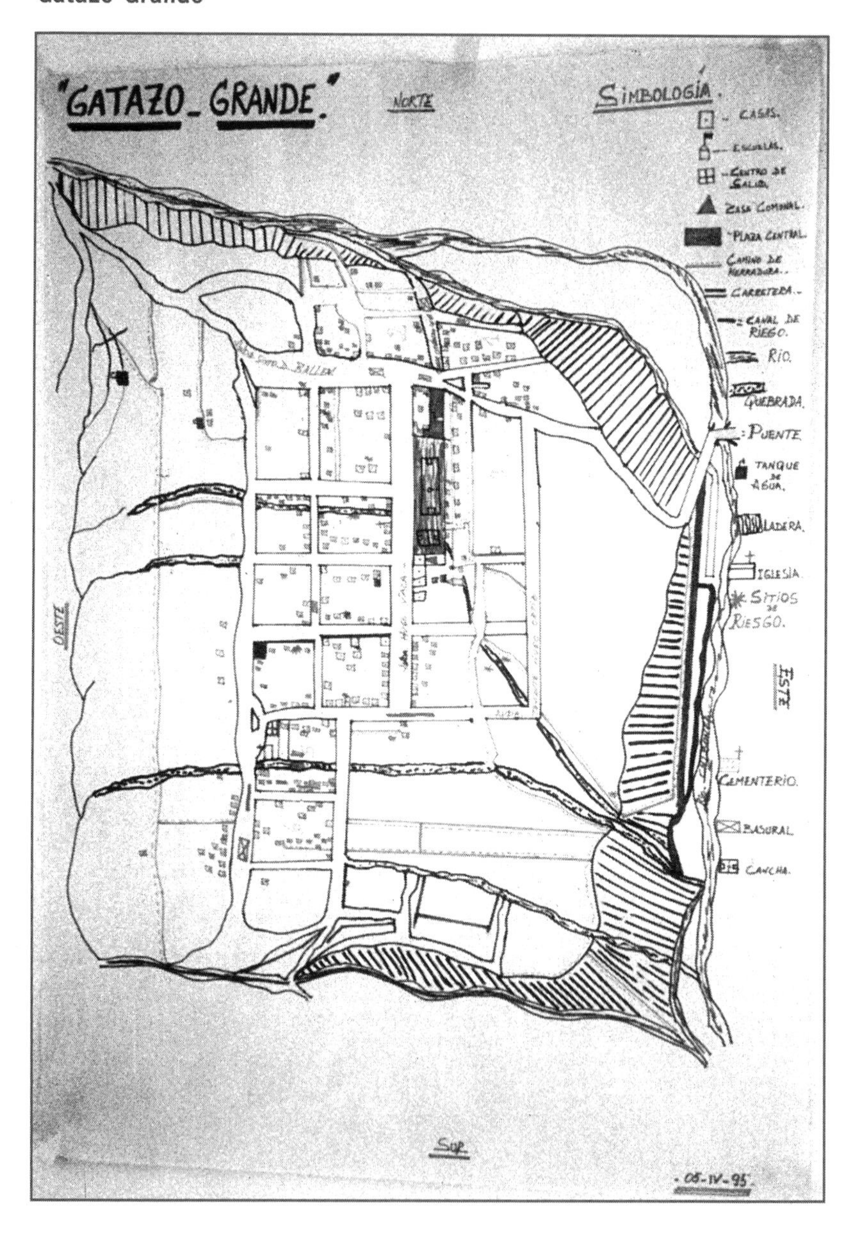

Comunidades Zona del Canal

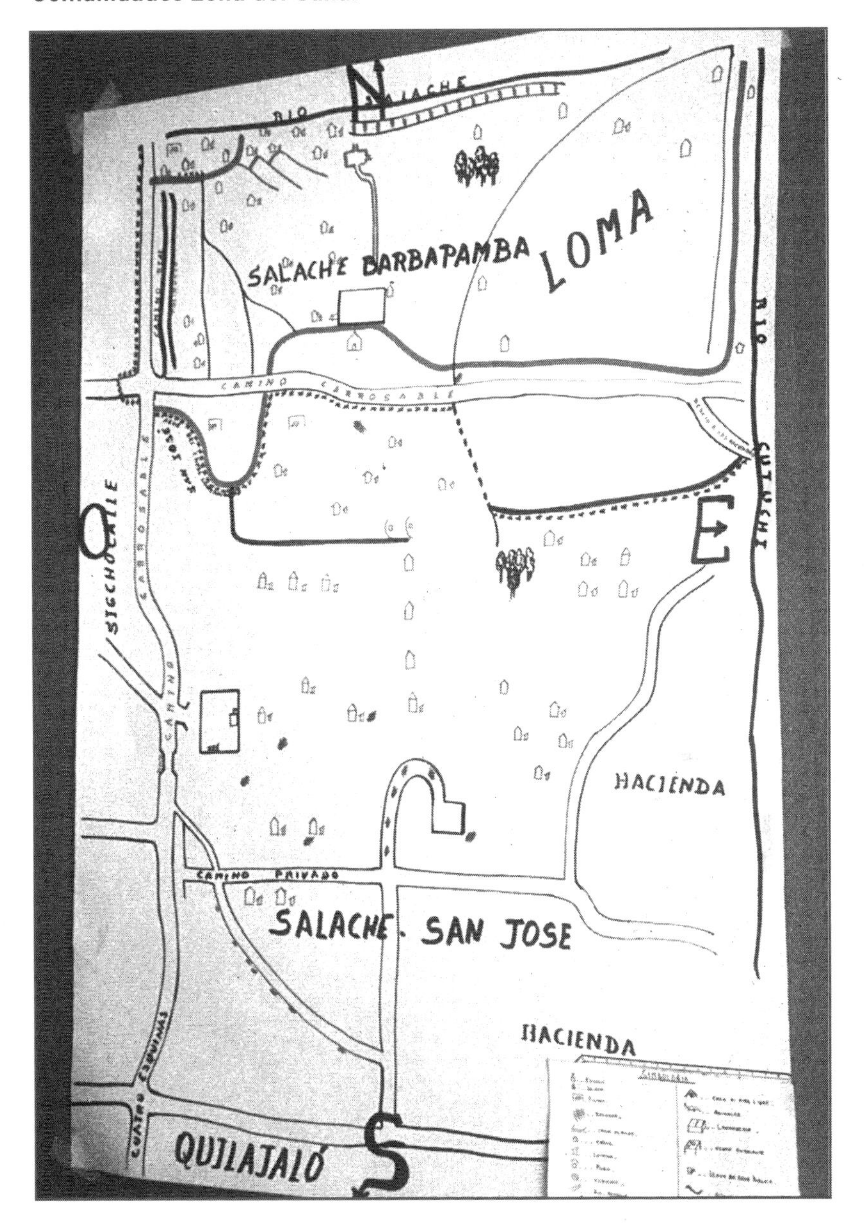

Alpamalag de la Cooperativa

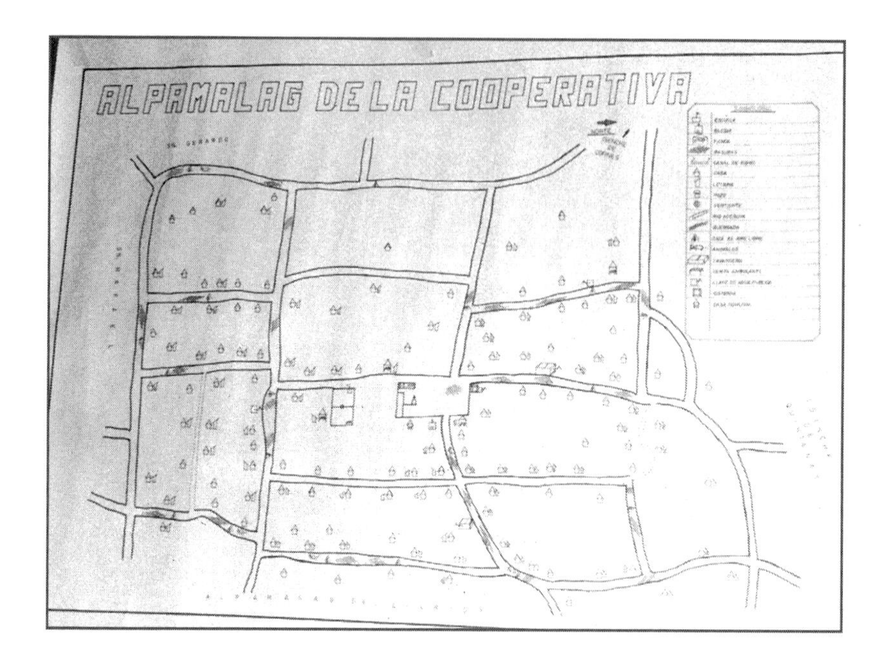

Searchable Databases and On-line Resources for Class Use

Link to Ten Minute Video of Site Available for Class Use

- CPI – A Model for the Prevention of Infectious Disease: youtube.com/watch?v=3T67TREownA

General Resources

- CDC Wonder: wonder.cdc.gov
- Cholera and Other Resources on the Disease: www.cdc.gov/cholera/index.html
- Cholera Global Situation and Trends: www.who.int/gho/epidemic_diseases/cholera/en/
- Demographic and Health Surveys (DHS): www.dhsprogram.com/
- E-Learning Resources for Global Health Researchers from the NIH: www.fic.nih.gov/Global/Pages/training-resources.aspx
- Global Atlas of Infectious Diseases by the World Health Organization: apps.who.int/globalatlas/
- Global Health at the Centers for Disease Control and Prevention (CDC): www.cdc.gov/globalhealth/index.html
- Global Health News from NPR: www.npr.org/sections/global-health/

Community Participatory Involvement: A Sustainable Model for Global Public Health by Linda M. Whiteford and Cecilia Vindrola-Padros, 171–174.

- Global Health Observatory Data Repository (by topic): apps.who.int/gho/data/?theme=main
- Global Health Resources at the National Institutes of Health – Fogarty International Center: www.fic.nih.gov/Global/Pages/default.aspx
- Infectious Disease Resources at the National Foundation for Infectious Diseases: www.nfid.org/links/id
- Infectious Disease Resources from Public Health: www.publichealth.org/resources/infectious-disease/
- Information on Cholera from the World Health Organization: www.who.int/mediacentre/factsheets/fs107/en/
- National Center for Infectious Diseases – Resources for Students: www.cdc.gov/ncidod/student.htm
- Publications, Data, & Statistics of Cholera from the CDC: www.cdc.gov/cholera/publications.html
- Regional Coalition for Water and Sanitation to Eliminate Cholera in Hispanola: www.paho.org/coleracoalicion/
- Resource Centre at the World Health Organization: www.who.int/hrh/resources/en/
- Supercourse: Epidemiology, the Internet, and Global Health – a Collaboration of the World Health Organization and the University of Pittsburgh: www.pitt.edu/~super1/

Activities

- Ethical Challenges in Short Term Global Health Training: www.ethicsandglobalhealth.org/
- Field Guide for Developing at Risk Communication Strategy: From Theory to Action: www1.paho.org/cdmedia/riskcommguide/
- Global Health Case Studies from the Center for Global Development: www.cgdev.org/page/case-studies
- Global Health Educational Modules from the Consortium of Universities for Global Health: cugh.org/resources/educational-modules#Global%20Health
- Global Trends (from PBS NOVA): www.pbs.org/wgbh/nova/earth/global-trends-quiz.html

- The Earth in Peril (from PBS NOVA): www.pbs.org/wgbh/nova/earth/earth-peril.html
- Rx for Survival: A Global Health Challenge (by PBS): www.pbs.org/wgbh/rxforsurvival/series/teachers/
    - Rx for Survival: Deadly Disease: www.pbs.org/wgbh/rxforsurvival/series/diseases/index.html

Interactive Websites

- CDC Flu Tracker: www.cdc.gov/flu/weekly/fluviewinteractive.htm
- Emergency Disaster Information Service: hisz.rsoe.hu/alertmap/index2.php
- Google Flu Trends: www.google.org/flutrends/
- Health Map: www.healthmap.org/en/

Videos

- The Story of Cholera by Global Health Media: globalhealthmedia.org/videos/cholera-portfolio/
- Top Documentary Films (free to view), topic – Health: topdocumentaryfilms.com/category/health/

Other

- Antibiotic Treatment for Cholera Information from the CDC: www.cdc.gov/cholera/treatment/antibiotic-treatment.html
- Cholera on Partners in Health (PIH): www.pih.org/priority-programs/cholera
- Community Health Worker Cholera Training from Partners in Health: www.pih.org/library/community-health-worker-cholera-training
- Diseases and Conditions Database from the CDC: www.cdc.gov/DiseasesConditions/
- Free Courseware from Johns Hopkins School of Public Health: ocw.jhsph.edu/
- FRONTLINE Rough Cut – Profile on Ecuador and Related Links: www.pbs.org/frontlineworld/rough/2007/06/ecuador_healthlinks.html

- Global Alert and Response (GAR) at the World Health Organization: www.who.int/csr/en/
- Infectious Diseases Society of America: www.idsociety.org/Index.aspx
- Public Health Image Library from the CDC: phil.cdc.gov/Phil/home.asp
- Stop Cholera Project: www.stopcholera.org/
- United States Department of Human Health and Human Services – Global Health: www.globalhealth.gov/

References

Altman, David G.

1995 "Sustaining Interventions in Community Systems: On the Relationship between Researchers and Communities," Health Psychology 14: 526–536.

Armitage, Christopher and Mark Conner

2000 "Social Cognition Models and Health Behavior: A Structured Review," Psychology and Health 15: 173–189.

2001 "Efficacy of the Theory of Planned Behavior: A Meta-Analytic Review," British Journal of Social Psychology 40: 471–499.

Baer, Hans

2011 Medical Pluralism: An Evolving and Contested Concept in Medical Anthropology. *In* A Companion to Medical Anthropology, Merrill Singer and Pamela I. Erickson, eds. Pp. 405–423. Malden, MA: Wiley-Blackwell.

Baer, Hans with Merrill Singer and Ida Susser

2003 Medical Anthropology and the World System. Westport, CT: Praeger.

Baer, Hans and Merrill Singer

2009 Global Warming and the Political Ecology of Health: Emerging Crises and Systemic Solutions. Walnut Creek, CA: Left Coast Press, Inc.

Bahamonde Harvez, C. and V. Stuardo Avila

2013 "La Epidemia de Colera en América Latina: Reemergencia y Morbimortalidad. Revista Panamericana Salud Publica," 33(1): 40–46.

Barragan Arenas, Italo and Edgardo Torres

1991 Guía para la Prevención y Control del Cólera. Quito, Ecuador: Organización Panamericana de la Salud.

Bartholomew, L. Kay, Guy S. Paracel, Gergo Kok, Nell H. Gottleib, and Maria E. Fernandez

2011 Planning Health Promotion Programs: An Intervention Mapping Approach. San Francisco: Jossey-Bass.

Barua, Dhiman

1972 "The Global Epidemiology of Cholera in Recent Years," Proceedings of the Royal Society of Medicine 65: 423–428.

Borda, Orlando Fals

1990 The Application of Participatory-Action Research in Latin America. *In* Globalization, Knowledge, and Society: Readings from International Sociology, M. Albrow and E. King, eds. Pp. 79–96. Newbury Park, CA: Sage.

Bourdieu, Pierre

2001 Masculine Domination. Palo Alto, CA: Stanford University Press.

Boutron, Isabelle with David Moher, Douglas Altman, Kenneth Schulz, Philippe Ravaud and the CONSORT Group

2008 "Methods and Processes of the CONSORT Group: Example of an Extension for Trials Assessing Nonpharmacologic Treatments," Annals of Internal Medicine 148(4): W60–W66.

Briggs, Charles

2001 "Modernity, Cultural Reasoning, and the Institutionalization of Social Inequality: Racializing Death in a Venezuelan Cholera Epidemic," Comparative Studies in Society and History 43(4): 665–700.

2005 "Communicability, Racial Disclosure, and Disease," Annual Review of Anthropology 34: 269–291.

2008 "A Framework for Integrated Environmental Health and Impact Assessment of Systemic Risks," Environmental Health 7(61): 1–17.

Briggs, Charles and Clara Mantini-Briggs

2003 Stories in the Time of Cholera: Racial Profiling during a Medical Nightmare. Berkeley: University of California Press.

Carpenter, Christopher

2010 "A Meta-Analysis of the Effectiveness of Health Belief Models Variables in Predicting Behavior," Health Communication 25: 661–669.

Castro, Arachu and Merrill Singer

2004 Unhealthy Health Policy: A Critical Anthropological Examination. Walnut Creek, CA: AltaMira Press.

Centers for Disease Control (CDC)

2011 Cholera in Haiti: One Year Later. www.cdc.gov/haiticholera/haiti_cholera.htm, last accessed April 9, 2015.

2013 Cholera-Vibrio cholerae infection. www.cdc.gov/cholera/general/, last accessed May 24, 2015.

Chevallier, E. with A. Grand and Azais, J.

2004 "Spatial and Temporal Distribution of Cholera in Ecuador between 1991 and 1996," European Journal of Public Health 14(3): 274–279.

Conner, Mark with Russell Bell and Paul Norman

2002 "The Theory of Planned Behavior and Healthy Eating," Health Psychology 21(2): 194–201.

Conner, Mark and Paul Norman

2005 Predicting Health Behavior: A Social Cognition Approach. *In* Predicting Health Behavior, Mark Conner and Paul Norman, eds. Pp. 1–27. Berkshire, UK: Open University Press.

Coreil, Jeannine, ed.

2010 Social and Behavioral Foundations of Public Health. Thousand Oaks, CA: Sage.

Coreil, Jeanine and Linda Whiteford and Diego Salazar

2000 The Household Ecology of Disease Transmission: Dengue Fever in the Dominican Republic. *In* The Anthropology of Infectious Disease: International Health Perspectives, Peter J. Brown and Marcia C. Inhorn, eds. Pp. 145–171. Amsterdam, Netherlands: Routledge.

Coupal, Francoise

1995 Participatory Project Design: Its Implications for Evaluation: A Case Study from El Salvador. www.mosaic-net-intl.ca/elsalvador.shtml, last accessed November 12, 2007.

Cox, Stephen and Sheldon Annis

1988 Community Participation in Rural Water Supply. *In* Direct to the Poor: Grassroots Development in Latin America, S. Annis and P. Hakim, eds. Pp. 65–72. Boulder, CO: Lynne Rienner.

Creamer G. with N. Leon, M. Kenber, P. Samaniego and G. Buchholz

1999 "Efficiency of Hospital Treatment in Ecuador," Revista Panamericana Salud Publica 5(2): 77–87.

De Koning, K. and M. Martin

1996 Participatory Research in Health: Issues and Experiences. London: Zed Books.

DiPrete Brown, L. and E. Hurtado.

1992 Development of a Behavior-Based Monitoring System for the Health Education Component of the Rural Water and Health Project CARE/Guatemala. WASH Field Report No. 364.

Fals Borda, Orlando

1990 The Application of Participatory-Action Research in Latin America. *In* Globalization, Knowledge, and Society: Readings from International Sociology, M. Albrow and E. King, eds. Pp. 79–96. Newbury Park, CA: Sage.

Farmer, Paul

1994 Uses of Haiti (Vol. 3.) Monro, ME: Common Courage Press.

1997 "Letters from Haiti," AIDS Clinical Care 9(11): 83–85.

1999 Infections and Inequalities: The Modern Plagues. Berkeley: University of California Press.

2003 Pathologies of Power. Berkeley: University of California Press.

2004 "An Anthropology of Structural Violence," Current Anthropology 45(3): 305–325.

2005 "Global AIDS: New Challenges for Health and Human Rights," Perspectives in Biology and Medicine 48(1): 10–16.

Farmer, Paul and Arachu Castro

2004 Pearls of the Antilles? Public Health in Haiti and Cuba. *In* Unhealthy Health Policy: A Critical Anthropological Examination, Arachu Castro and Merrill Singer, eds. Pp. 3–28. Lanham, MD: AltaMira Press.

Faruque, Shah, Nityananda Chowdhury, M. Kamruzzaman, Q. Shafi Ahmad, A. S. G. Faruque, M. Abdus Salam, T. Ramamurthy, G. Balakrish Nair, Andrej Weintraub, and David A. Sack

2003 "Reemergence of Epidemic Vibrio cholerae O139, Bangladesh," Emerging Infectious Diseases 9(9): 1116–1122.

Fassin, Didier

2009 A Violence of History: Accounting of AIDS in Post-Apartheid South Africa. *In* Global Health in Times of Violence. Barbara Rylko-Bauer, Linda Whiteford, and Paul Farmer, eds. Pp. 113–136. Santa Fe, NM: School of Advanced Research Press.

Fisher, Jeffrey with William Fisher, K. Rivet Amico and Jennifer Harman

2006 "An Information-Motivation-Behavioral Skills Model of Adherence to Antiretroviral Therapy," Health Psychology 25(4): 462–473.

Freire, Paulo

1970 Pedagogy of the Oppressed. New York: Bergman, Herder and Herder.

Frelick, G., Jennings, L. and P. Haggerty

1993 Preparation for Conducting a Second Training of Trainers Workshop and Producing a Training Guide for the Development of a Hygiene Education Program.WASH Field Report No. 83. Arlington, VA: WASH Project.

Gabastou, J. M. with C. Pesantes, S. Escalante, Y. Narvaez, E. Vela, L. Garcia, D. Zabala, and Z. Yadon

2002 "Caracteristicas de la Epidemia de Colera de 1998 en Ecuador, Durante el Fenomeno del Nino," Revista Panamericana Salud Publica 12(3): 157–164.

Galtung, Johan

1969 "Violence, Peace and Peace Research," Journal of Peace Research 6(3): 167–191.

Glanz, Karen and Donald Bishop

2010 "The Role of Behavioral Science Theory in Development and Implementation of Public Health Interventions," Annual Review of Public Health 31: 399–418.

Godoy Paiz, Paula

2007 Everyday Violence: Struggle and Resilience among Guatemalan Women. www.mcgill.ca/trauma-global health/print/26, last accessed September 22, 2012.

Green, L. W. with M. A. George, M. Daniel, et al.

1995 Study of Participatory Research in Health Promotion: Review and Recommendations for the Development of Participatory Research in Health Promotion in Canada. Vancouver, British Columbia: Royal Society of Canada.

Guttmacher, Sally with Patricia Kelly and Yumary Ruiz-Janecko

2010 Community-Based Health Interventions: Principles and Applications. San Francisco: Jossey-Bass.

Hall, Budd L.

1992 "From Margins to Center? The Development and Purpose of Participatory Research," American Sociological Review 23: 15–28.

Hardeman, Wendy with Marie Johnston, Derek Johnston, Debbie Bonetti, Nicholas Wareham, Ann Louise Kinmonth

2002 "Applications of the Theory of Planned Behavior Change Interventions: A Systematic Review," Psychology and Health 17(2): 123–158.

Hulscher, M. E. with M. G. Laurant and R. P. Grol

2003 "Process Evaluation on Quality Improvement Interventions," Quality Safety Health Care 12: 40–46.

Husereau, Don with Michael Drummond, Stavros Petrou, Chris Carswell, David Moher, Dan Greenberg, Federico Augustovski, Andrew Briggs, Josephine Mauskopf, Elizabeth Loder, and CHEERS Task Force

2013 "Consolidated Health Economic Evaluation Reporting Standards (CHEERS) Statement," BMC Medicine 11: 80–86.

IRIN

2006 Sudan: Cholera death toll in south rises to 238. www.irinnews.org/report/58524/sudan-cholera-death-toll-in-south-rises-to-238, last accessed May 24, 2015.

Israel, Barbara with A. Schulz, E. Parker, and A. Becker

1998 "Review of Community Based Research: Assessing Partnership Approaches to Improve Public Health," Annual Review of Public Health 19: 173–202.

Johnson, Steven

2006 The Ghost Map: The Story of London's Most Terrifying Epidemic – And How It Changed Science, Cities, and the Modern World. New York: Riverhead Books.

Kleinman, Arthur and Peter Benson

2006 "Anthropology in the Clinic: The Problem of Cultural Competency and How to Fix It," PLos Medicine 3(10): e294.

Lee, Kelly and Richard Dodgson

2000 "Globalization and Cholera: Implications for Global Governance," Global Governance 6(2): 213–236.

Lee, P. with N. Krause and C. Goetchius

2003 Participatory Action Research with Hotel Room Cleaners: From Collaborative Study to the Bargaining Table. *In* Community-Based Participatory Research for Health, M. Minkler and N. Wallerstein, eds. Pp. 390–404. San Francisco: Jossey-Bass.

Lippke, Sonia and Jochen P. Ziegelmann

2006 "Understanding and Modeling Health Behavior: The Multi-Stage Model of Health Behavior Change," Journal of Health Psychology 11(1): 37–50.

Loyd, Jenna M.

2009 "'A Microscopic Insurgent': Militarization, Health, and the Critical Geographies of Violence," Annals of the Association of American Geographers 99(5): 863–873.

Malavade, S. with A. Narvaez, A. Mitra, T. Ochoa, E. Naik, M. Sharma, S. Galwankar, M. Breglia and R. Izurieta

2011 "Cholera in Ecuador: Current Relevance of Past Lessons Learnt," Journal of Global Infectious Disease 3(2): 189–194.

McElroy, Ann and Patricia Townsend

2009 Medical Anthropology in Ecological Perspective. Boulder, CO: Westview Press.

McLeroy, Kenneth with Daniel Bibeau, Allan Steckler and Karen Glanz

1988 An Ecological Perspective on Health Promotion Programs," Health Education Quarterly 15(4): 351–377.

Michie, Susan and Charles Abraham

2004 "Interventions to Change Health Behaviours: Evidence-based or Evidence-inspired?" Psychology and Health 19(1): 29–49.

Minkler, Meredith

2004 "Community-Based Research Partnerships: Challenges and Opportunities," Journal of Urban Health 82(2): ii3–ii12.

Misovich, Stephen, Todd Martinez, Jeffrey Fisher, Angela Bryan and Nicole Catapano

2003 Predicting Breast Self-examination: A Test of the Information-Motivation-Behavioral Skills Model," Journal of Applied Social Psychology 33(4): 775–790.

MMWR

Sept 3, 2004 Cholera epidemic associated with raw vegetables-Lusaka, Zambia, 2003-2004. MMWR Weekly 53(34): 783–786. www.cdc.gov/mmwr/preview/mmwrhtml/mm5334a2.htm

Montgomery, Maggie A., and Menachem Eimelech

2007 "Water and Sanitation in Developing Countries: Including Health in the Equation," Environmental Science and Technology 41: 16–24.

MSF

2005 MSF reacts to new cholera outbreak in Burundi. www.msf.org/article/msf-reacts-new-cholera-outbreak-burundi, last accessed May 24, 2015.

Nalin, David R. and Richard A. Cash

1971 "Oral or Nasogastric Maintenance Therapy in Pediatric Cholera Patients," The Journal of Pediatrics 78(2): 355–358.

Narkevich, M. I., G. G. Onischenko, J. M. Lomov, E.A. Moskvitina, L. S. Podosinnikova, and G. M. Medinsky

1993 "The Seventh Pandemic of Cholera in the USSR, 1961–1989," Bulletin of the World Health Organization 71(2): 189–198.

National Public Radio

2011 Paul Farmer Examines Haiti 'After the Earthquake.' www.npr.org/2011/07/12/137762573/paul-farmer-examines-haiti-after-the-earthquake, last accessed November 18, 2014.

Nations, Marilyn K. and Cristina M. G. Monte

1996 "'I'm Not Dog, No!': Cries of Resistance Against Cholera Control Campaign," Social Science & Medicine 43(6): 1007–1024.

Nichter, Mark

2008 Global Health: Why Cultural Perceptions, Social Representations, and Biopolitics Matter. Tucson: University of Arizona Press.

Ogrinc, G. with S. E. Mooney, C. Estrada, T. Foster, D. Goldmann, L. Hall, M. Huizinga, S. Liu, P. Mills, J. Neily, W. Nelson, P. Pronovost, L. Provost, L. Rubenstein, T. Speroff, M. Splaine, R. Thomson, A. Tomolo, and B. Watts

2008 "The SQUIRE (Standards for Quality Improvement Reporting Excellence) Guidelines for Quality Improvement Reporting: Explanation and Elaboration," Quality Safety Health Care 17: i13–i32.

Ovretveit, John

2014 Evaluating Improvement and Implementation for Health. Berkshire, UK: Open University Press.

Page, J. Bryan and Merrill Singer

2010 Comprehending Drug Use: Ethnographic Research at the Social Margins. New Brunswick, NJ: Rutgers University Press.

Panter-Brick, Catherine, Sian Clarke, Heather Lomas, Margaret Pinder and Steve Lindsay

2006 "Culturally Compelling Strategies for Behavior Change: A Social Ecology Model and Case Study in Malaria Prevention," Social Science & Medicine 62: 2810–2825.

Prochaska, James O. and Carlo C. DiClemente

1994 The Transtheoretical Approach: Crossing Traditional Boundaries of Therapy. Ann Arbor: University of Michigan Press.

Prochaska, James, Colleen Redding, Lisa Harlow, Joseph Rossi and Wayne Velicer

1994 The Transtheoretical Model of Change and HIV Prevention: A Review," Health Education Quarterly 21(4): 471–486.

Rhodes, Scott with Robert Malow and Christine Jolly

2010 "Community-based Participatory Research (CBPR): A New and Not-so-New Approach to HIV/AIDS Prevention, Care, and Treatment," AIDS Education and Prevention 22(3): 173–183.

Rosenstock, I. M.

1966 "Why People Use Health Services," Milbank Memorial Fund Quarterly 44: 94–127.

Rylko-Bauer, Barbara and Paul Farmer

2002 "Managed Care or Managed Inequality? A Call for Critiques of Market-Based Medicine," Medical Anthropology Quarterly 16(4): 476–502.

Rylko-Bauer, Barbara with Linda Whiteford and Paul Farmer, eds.

2009 Global Health in Times of Violence. Santa Fe, NM: School for Advanced Research Press.

Saksvik, Per Oystein with Kjell Nytro, Carla Dahl-Jorgensen and Aslaug Mikkelsen

2002 "A Process Evaluation of Individual and Organizational Occupational Stress and Health Interventions," Work & Stress 16(1): 37–57.

Sanchez, J. L. and D. N. Taylor

1997 "Cholera," Lancet 349(9068): 1825–1830.

Schensul, J. J. and M. D. LeCompte

2010 Ethnographer's Toolkit: 7-volume paperback boxed set, 2nd edition. Lanaham, MD: AltaMira Press.

Scheper-Hughes, Nancy

1992 Death Without Weeping: Violence in the Everyday Life in Brazil. Berkeley: University of California Press.

Schuller, M.

2006 Homing Devices: The Poor as Targets of Public Housing Policy and Practice. Lanham, MD: Lexington Books.

Schulz, A. J. with B. A. Israel, S. M. Selig, I. S. Bayer and C. B. Griffin

1998 "Development and Implementation of Principles for Community-based Research in Public Health," *In* Research Strategies for Community Practice. R. H. MacNair, ed. Pp. 83–110. New York: Haworth Press.

Sidani, Souraya and Carrie Jo Braden

2011 Design, Evaluation, and Translation of Nursing Interventions. West Sussex, UK: John Wiley and Sons.

Singer, Merrill and Pamela Erickson, eds.

2011 A Companion to Medical Anthropology. Hoboken, NJ: Wiley-Blackwell.

Singer, Merrill and Scott Clair

2003 "Syndemics and Public Health: Reconceptualizing Disease in Bio-Social Context," Medical Anthropology Quarterly 17(4): 423–441.

Singer, Merrill with Freddie Valentin, Hans Baer and Zhongke Jia

1992 "Why Does Juan Garcia Have a Drinking Problem? The Perspective of Critical Medical Anthropology," Medical Anthropology 14: 77–108.

Smith, S. E. with D. G. Willms and N.A. Johnson

1997 Nurtured by Knowledge: Learning to do Participatory Action-Research. New York: Apex Press.

Speck, Robert

1993 Cholera. *In* Cambridge World History of Human Disease, K. F. Kiple, ed. Pp. 642–649. Cambridge, UK: Cambridge University Press.

Stetler, Cheryl with Marcia Legro, Carolyn Wallace, Candice Bowman, Marylou Guihan, Hildi Hagedorn, Barbara Kimmel, Nancy Sharp, and Jeffrey Smith

2006 "The Role of Formative Evaluation in Implementation Research the QUERI Experience," Journal of General Internal Medicine 21: S1–S8.

Stokols, Daniel

1992 "Establishing and Maintaining Health Environments: Toward a Social Ecology of Health Promotion," American Psychologist 47(1): 6–22.

Taylor, Janelle S.

2003 "Confronting 'Culture' in Medicine's 'Culture of No Culture,'"Academic Medicine 78(6): 555–559.

Thomas, David R.

2006 "A General Inductive Approach for Analyzing Qualitative Evaluation Data," American Journal of Evaluation 27(2): 237–246.

Trinchero, Hugo and Juan Martin Leguizamon

1995 "The Cholera Emergence in Argentina and the Construction of Ethnic Stigmas Towards Indigenous People on the Border with Paraguay and Bolivia," Bulletin of the International Committee on Urgent Anthropological and Ethnological Research 37: 99–113.

Trostle, James A.

2005 Epidemiology and Culture. Cambridge: Cambridge University Press.

UNESCO

1997 Adult learning and the challenges of the 21st Century. Fifth International Conference on Adult Education. Hamburg, Germany: UNESCO, Institute of Education.

United Nations

2014 Haiti Cholera Response. United Nations in Haiti Factsheet. www.un.org/.../haiti/Cholera_UN_Factsheet_September_2014.pdf, last accessed May 22, 2015.

University of Notre Dame

2012 "Paul Farmer and Rev. Gustavo Gutiérrez to take part in campus dialogue." newsinfo.nd.edu/news/26883-paul-farmer-and-fr-gustavo-gutirrez-to-take-part-in-campus-dialogue/, last accessed July 16, 2012.

Walker, Kimberly and Richard Jackson

2015 "The Health Belief Model and Determinants of Oral Hygiene Practices and Beliefs in Preteen Children: A Pilot Study," Pediatric Dentistry 37(1): 40–45.

Wallerstein, Nina and Bonnie Duran

2006 "Using Community-Based Participatory Research to Address Health Disparities," Health Promotion Practice 7(3): 312–323.

2010 "Community Based Participatory Contributions to Intervention Research: The Intersection of Science and Practice to Improve Health Equity," American Journal of Public Health S1(100): S40–S46.

Waldman, R. J., E. D. Mintz, and H. E. Papowitz

2013 "The Cure for Cholera-Improving Access to Safe Water and Sanitation," The New England Journal of Medicine 368(7): 592–594.

Walter, Nicholas, Philippe Bourgois, and H. Margarita Loinaz

2004 "Masculinity and Undocumented Labor Migration: Injured Latino Day Laborers in San Francisco," Social Science & Medicine 59: 1159–1168.

Washington Post

2014 "Ebola's lessons, painfully learned at great cost in dollars and human lives." Washington Post, December 28.

Weber, J. T., E. Mintz, R. Canizares, A Semiglia, I. Gomez, R. Sempertegui, A. Davila, K. Greene, N. Puhr, D. Cameron, F. Tenover, T. Barrett, N. Bean, C. Ivey, R. Tauxe, and P. Blake

1994 "Epidemic Cholera in Ecuador: Multidrug-resistance and Transmission by Water and Seafood," Epidemiology & Infection 112: 1–11.

Whiteford, Linda M.

1997 The Ethnoecology of Dengue Fever. Special Issues: "Knowledge and Practice in International Health" P. Stanley Yoder, Guest Editor. Medical Anthropology Quarterly 11(2): 202.

1999 Environmental Health Project: Community Involvement in the Management of Environmental Pollution (CIMEP) Guidelines, Manual. U.S. Agency for International Development.

2003 Cholera and the Anthropological Contributions to its Understanding. *In* Encyclopedia of Medical Anthropology. Pp. 305–311. New Haven, CT: HRAF Press.

2009a The Medical Ecology of Cholera in Ecuador. *In* Medical Anthropology in Ecologcal Perspective, Ann McElroy and Patricia Townsend, eds. Pp. 375–382. Boulder, CO: Westview Press.

2009b Approaches to Policy and Advocacy. *In* Social and Behavioral Foundations of Public Health, J. Coreil, ed. Pp. 311–324. Thousand Oaks, CA: Sage.

2010 Approaches to Policy and Advocacy. *In* Social and Behavioral Foundations of Public Health. Jeannine Coreil, ed. Pp. 311–324. Thousand Oaks, CA: Sage.

Whiteford, L. M. and L. Bennett

2005 Applied Anthropology and Health and Medicine. *In* Applied Anthropology: Domains of Application, Satish Kedia and John van Willigen, eds. Pp. 119–149. Westport, CT: Greenwood Press.

Whiteford, L. M. and J. Coreil

1997 The Household Ecology of Disease Transmission: Dengue Fever in The Dominican Republic. *In* Anthropology and Infectious Disease, Peter Brown and Marsha Inhorn, eds. Pp. 143–172. Westport, CT: Greenwood Press.

Whiteford, Linda and Carmen Laspina

1996 Evaluation of Cholera Control Activities: The BACA Project. CDM Environmental Health Project, Arlington ,VA.

Whiteford, Linda, Carmen Laspina, and Mercedes Torres

1996 Cholera Prevention in Ecuador: Community-based Approaches for Behavior Change. Activity Report No.19, Environmental Health Project, Arlington, VA.

Whiteford, Linda and Robert T. Trotter
2008 Anthropological Ethics for Research and Practice. Long Grove, IL: Waveland Press.

Whiteford, Linda and S. Whiteford, eds.
2005 Casualties in the Globalization of Water: A Moral Economy of Perspective. *In* Globalization, Water and Health: Resources in Times of Scarcity. Pp. 25–45. Santa Fe, NM: School of American Research Press.

Williams, Lynn, with Susan Rasmussen, Adam Kleczkowski, Savi Maharaj, and Nicole Cairns
2015 "Protection Motivation Theory and Social Distancing Behaviour in Response to a Simulated Infectious Disease Epidemic," Psychology, Health, and Medicine 2: 1–6 [Epub ahead of print].

Wynn, Barbara, with Arindam Dutta, and Martha Nelson
2006 Challenges in Program Evaluation of Health Interventions in Developing Countries. Santa Monica, CA: Rand Corporation.

World Health Organization (WHO)
2003 WHO Report in Global Surveillance of Epidemic-prone Infectious Diseases-Cholera. Geneva: World Health Organization.

2006 Cholera in South Sudan, Update.

2012 Cholera Fact Sheet 107. World Health Organization.

2013 "Cholera, 2012," Weekly Epidemiological Record 31(88): 321–336.

2014 Cholera fact sheet N107. www.who.int/mediacentre/fact-sheets/fs107/en/, last accessed May 24, 2015.

2015 Global Health Observatory Data Repository. apps.who.int/gho/data/node.main.175?lang=en, last accessed May 28, 2015.

Yacoob, May, Dan O'Brien, and Rick Henning
1989 CARE Indonesia: Increasing Community Participation and Developing a Basic Strategy for Hygiene Education in Rural Water and Sanitation Programs. WASH Field Report No. 284. Arlington, VA: WASH Project.

Yacoob, May, Michael Carroll, Javier Chan, Fleming Helgaard, Santos Mahung, Anthony Nicasio, Jorge Polanco, Jerry VanSanti, Francis Westby, and Alan Wyatt.

1991 Improved Productivity Through Better Health (IPTBH) Project Assessment. WASH Field Report 356. Arlington, VA: WASH Project.

Yacoob, May, Santos Mahung, Michael Carroll, and Fleming Helgaard

1992 Program Planning Workshop for the Improved Productivity through Better Health Project. WASH Field Report No. 365. Arlington, VA: WASH Project.

Yacoob, May, G. Brantly, and Linda Whiteford

1994 Public Participation in Urban Environmental Management, WASH Technical Report No 90.

Index

About the Authors

Linda M. Whiteford is a Professor of Anthropology and Co-Director of the World Health Organization (WHO) Collaborating Center at the University of South Florida, where she also has served as Vice Provost for Academic Program Development and Review, Associate Vice President for Global Strategies, and Associate Vice President for Strategic Initiatives. Former chair of the Department of Anthropology, she helped develop the USF Patel School of Global Sustainability, the Global Academic Partners Program, the USF Office of Sustainability, USF World, and the USF Office of Community Engagement. She is an internationally recognized researcher, lecturer, and author; a medical anthropologist who has consulted for the World Bank, the World Health Organization, the Pan American Health Organization; and an advisor on global health policy. Her most recent book, *Global Health in Times of Violence,* is co-edited with Paul Farmer (Harvard University) and Barbara Rylko-Bauer (Michigan State University).

Cecilia Vindrola-Padros is a Research Associate in the Department of Applied Health Research at University College London. She obtained a PhD in Medical Anthropology from the University of South Florida. She currently works as an embedded qualitative researcher in various hospitals in the United Kingdom, designing and implementing evaluations and developing collaborative strategies with clinical teams to address issues affecting health service delivery. Among her recent publications are: "The Narrated, the Nonnarrated, and the Disnarrated: Conceptual Tools for Analyzing Narratives" (in *Qualitative Health Research*) and "Together, We Can Show You: Using Participant-Generated Visual Data in Collaborative Research" (*Collaborative Anthropologies*).